ELECT

MO

compiled by

M. H. BABANI B.Sc.(Eng.)

BERNARD BABANI (publishing) LTD
THE GRAMPIANS
SHEPHERDS BUSH ROAD
LONDON W6 7NF
ENGLAND

The compiler of this book wishes to express his most sincere thanks to "ELECTRONICS AUSTRALIA" the leading radio, T.V. and electronics magazine published in that continent and to Sungravure Syndication, their agents, for permission to use some of the material in this book which appeared originally as articles in that magazine.

I.S.B.N. 0 900162 59 7

First Published - April 1976
Reprinted – July 1979

Printed and Manufactured in Great Britain by
C. Nicholls & Co Ltd, Manchester M11 4AU

CONTENTS

The compiler of this book would like to express his most sincere thanks to the following manufacturers : Varta, Ever Ready and SAFT for allowing use of information from their catalogues.

SIMPLE MODEL TRAIN CONTROL

Here is a simple model train control for those enthusiasts who desire something more than the usual type of rheostat control. The controller gives better low speed performance and is fully overload protected, yet contains a minimum of components. Best of all, you don't need to be an electronic genius to construct it.

The conventional method of controlling a model train is by means of a rheostat (or variable resistor) in series with the supply voltage. This provides continuous control of the armature current and hence of motor torque.

While a very simple method, it has one notable disadvantage: the amount of series resistance in the supply circuit tends to give it a "constant current" characteristic, largely defeating the beneficial effect of back EMF within the motor. In other circumstances, when a motor slows down, back EMF falls and armature current rises, increasing the torque; conversely, when a motor speeds up, back EMF rises and armature current falls, reducing torque.

When the supply has a "constant current" characteristic, the ability of the motor to vary the torque is limited and a model loco, for example, becomes very sensitive to load variations. Low speed behaviour is particularly poor: the loco tends to slow down markedly on slight hills, and may in some cases stall.

Low speed behaviour of a model train can be improved considerably by using a variable supply having a low impedance characteristic. At any given setting it will appear to the loco as a "constant voltage" rather than a "constant current" source.

Fortunately, the problem of providing a variable, low impedance supply for a model train can be greatly simplified by the application of electronic techniques. For example, a transistor connected in emitter follower mode, as in Figure 1, can impart a low impedance characteristic to an otherwise normal variable supply.

In itself, Figure 1 is not a practical circuit. A power transistor with a current gain of about 20, supplying up to 1.5 amps to the train, would require a base current rising to 75 milliamps. To minimise regulation problems in the potentiometer supplying the base, it would be necessary to arrange a current through it of the order of 1 amp. This would require a very high wattage potentiometer, and would be inefficient in terms of power consumption. The life of the power transistor could be of short duration, since the effect of a short circuit would be catastrophic – and short circuits do occur in practical systems.

The first of these problems can be overcome by the use of a Darlington pair. Figure 2 shows the circuit involved. Using a general purpose tran-

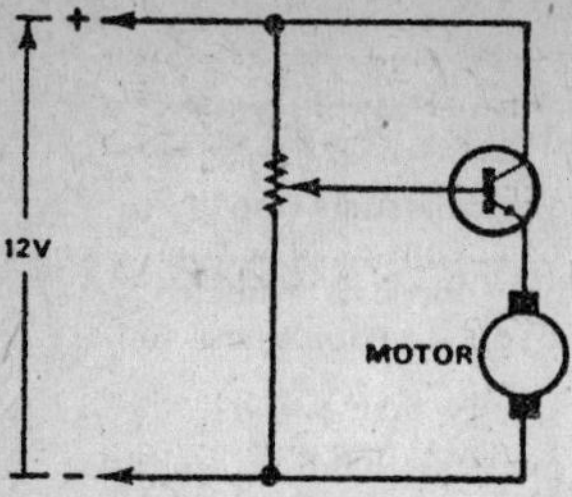

FIG. 1

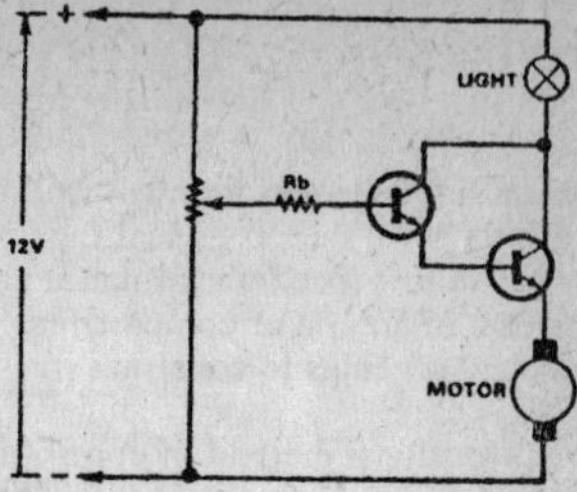

FIG. 2

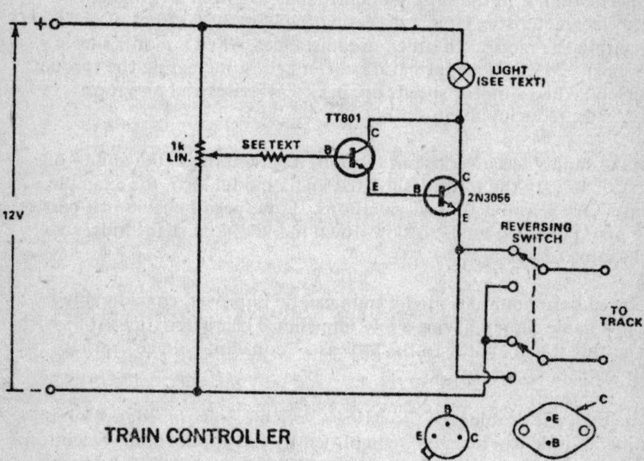

TRAIN CONTROLLER

FIG. 3

sistor in combination with a power transistor in a compound configuration, we can achieve a gain of 800 or so. The base current required can then be approximately 2 milliamps – a more practical figure. With a 12 volt supply rail, a 1k potentiometer is practical.

Protection against short circuits is provided by a filament lamp placed in the collector circuit and by a series base resistor.

The lamp functions as a non-linear resistance and, under normal running conditions, the current through it is only a small fraction of its rated full load current. The filament remains cold and the resistance low, so that only a small amount of power is diverted from the train.

When a short circuit occurs, the current increases, the filament becomes hot and it's resistance increases approximately by a factor of eight. This limits the increase in current, and hence the power dissipated in the transistor. Visible indication of the overload can be obtained by placing the light beneath a red bezel.

The purpose of the base resistor is to prevent the composite transistor from turning fully on or saturating before the speed control potentiometer reaches the full speed position. It also helps to limit the current during overload conditions. Its value is determined principally by the full load current of the lamp. With a 50 watt lamp, a 2.7k resistor is suitable, while for a 25 watt lamp, a 5.6k resistor is suitable.

The final circuit, with values, is shown in Figure 3.

The choice of lamp depends upon the power consumption of the train motor and any auxiliary equipment being supplied, such as carriage lights. For a single train only, it is possible to use a 25 to 30 watt 12V lamp, obtainable from most automotive spare parts dealers. A 27 watt lamp can be made by wiring the two filaments of a 21/6 watt stop/tail light in parallel.

If it is desired to "double head" trains, or to run auxiliary equipment, it will be necessary to use at least a 50 watt lamp, which can be obtained by using two stop/tail lights in parallel, or by using one filament of a headlight bulb. The former method is cheaper, but the latter may be more convenient. Those readers with older style cars may be able to use an old bulb which has one filament broken. For an exceptionally large or heavy load, it may be necessary to use both filaments of a headlight bulb in parallel.

No damage will be caused to the headlight bulb by using both filaments in parallel because it will only be energised fully for brief periods during overload conditions.

The controller requires a supply of 12V DC, rated at about 1.5A. This should be adequate provided that it is not overloaded for prolonged

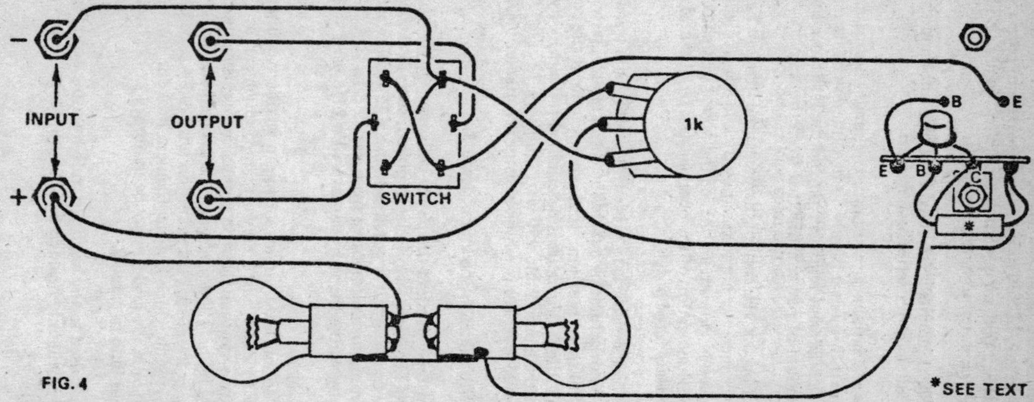

FIG. 4

The very simple wiring layout of the controller is shown above. Note that one of the lightbulbs is positioned underneath the bezel.

periods. It is not necessary for the supply to be smoothed; in fact a little AC ripple is helpful in overcoming the "stiction" inherent in most models. The normal commercial "train transformer/rectifier" is quite suitable as a supply.

Since the component and wiring layout is not critical, construction of the unit should cause no difficulties, even for inexperienced readers. A 5in x 2¼in aluminium box was used for a case, doubling as a heat-sink for the power transistor.

The input and output terminals were mounted on one end of the box, and the power transistor on the other end. This transistor must be insulated from the metalwork, using a mica washer and plastic bushes. Note that silicon grease should be used to thermally bond the case of the transistor to the box.
The mounting lug of a terminal strip was fixed to one of the mounting bolts of the transistor to provide the collector connection, the TT801 and the base resistor being then attached to the strip.

The overload bezel, reversing switch and speed control were mounted on the top of the box, and rubber feet on the bottom.

The lamp was a dual unit made from 21/6 watt car stop/tail lights. They were soldered together using tinned 18 gauge copper wire, which was then used to support them in position underneath the bezel, as shown in the photograph.
When the construction is completed, the unit can be tested. Connect a suitable power supply to the input terminals, making sure that the polarity is correct. Connect the output terminals to the tracks, and place a train on them. When the speed control is advanced, the train should run smoothly and it should be possible to vary its speed over a wide range.
While the train is at maximum speed, a screwdriver or some other suitable tool can be used to short the rails. The train should stop and the overload bezel should glow.

Part List

1 2N3055, RS276-2020, BD130.BDY139 power transistor with insulated mounting kit.
1 TT801 transistor.
1 2.7k ¼ watt resistor.
1 1k lin pot.
1 DPDT toggle switch.
1 Red bezel.
4 Terminals.
1 3-lug tap strip.
1 Lightbulb (see text).
1 Box (127 x 57 x 57 mm).
4 Rubber feet.
1 Knob.
Hookup wire, solder, screws, nuts, 18 gauge tinned copper wire.

MODEL TRAIN CONTROL WITH SIMULATED INERTIA

As a follow up to our train controller, here is a more elaborate version featuring simulated inertia, braking, and "kick" circuit to ensure reliable starting. Those who have already made the simple version will be in an ideal position to add these features to it.

An ideal model train would reproduce in miniature all the characteristics, problems and peculiarities of the real thing, be it an old coal-buring "Rocket" or a modern diesel-electric ore-train. Unfortunately, such models are unrealisable, and we have to be content with what is available, and more importantly, what we can afford. One area, however, in which improvements can be made is in the field of control.

The simplest method of control is to use a variable resistor in series with the supply voltage. This gives control of the armature current, and hence of the torque. At high speeds this gives quite acceptable control, since the circuit resistance, consisting of the armature resistance and the control resistance, is quite small, and the back emf is the major factor determining the armature current and, hence, the torque. This means that the speed of the train is substantially independent of load at high speeds.

Trouble occurs at low speeds, however. The control resistor is set for a large value, so there is a large resistance in the circuit. This means that the supply becomes a constant current supply, and so the back emf is not the major factor in determining the armature current. Hence the speed of the train varies quite markedly with the load, often stalling on gradients and curves.

Probably the most objectional feature of the simple resistance controller is its poor starting characteristics. It is almost impossible to start a train without it taking off like a rocket, no matter how skilfully the control knob is handled.

To understand why this is so, it is necessary to consider the following points:
(a) The torque of the motor is almost directly related to the current through the armature windings.
(b) A very much largerc urrent (torque) is needed to start the motor than is needed to keep it moving.
(c) The simple resistance controller tends to behave as a constant current type power supply; i.e., one in which the current through the armature tends to remain constant in spite of changes in the back emf which the armature generates.

Thus, when we provide enough current to start the train, we are automatically nominating this same value as the running current, due to the constant current characteristic. But this value is far too high for normal running speed, so the current has to be reduced – controller retarded – almost as soon as the train starts to move. Unless done skilfully, correction will either be inadequate or excessive; in the latter case stopping the train unnecessarily. Adding to this difficulty is the effects of the friction that occurs in the motor bearings and the contact problems inherent in transfering power from the rails to the train.

This is where the controller described in the April issue scores. It has a constant voltage characteristic, allowing the current to vary according to the motor's requirements, and as dictated by the back emf.

Anyone who has ever compared a real train with a model train would note one glaring difference. A real train, with a small power-to-weight ratio, cannot accelerate or brake rapidly. A model train, with a high power-to-weight ratio, is capable of ridiculous acceleration and braking performances. With some models, it is literally possible to stop a train in less than half the length of a carriage. (When this is done, mutters can usually be heard about firemen in fireboxes!) Of course, inertia can be simulated by being less heavy-handed on the throttle, but this technique is not suitable for automatic signalling systems, where trains have to be started and stopped automatically. What is obviously required is artificial inertia.

The problems and requirements just outlined inspired our new controller. The circuit is based on the Darlington pair emitter follower circuit described previously, to which we have added several features. The main one is the artificial inertia, the others being aimed mainly at getting the maximum benefit from the inertia circuit.

The inertia circuit itself can be switched out if required, since it has been found that precise shunting manoeuvres are sometimes more easily performed without it. The circuit has such good low speed characteristics that the operator can easily provide a realistic performance in these circumstances.

In conjunction with the inertia circuit we have a brake circuit. This provides added realism, in that the train may be allowed to simply coast after power is switched off, or brought to a stop more rapidly by applying the brake. While our circuit uses a simple push button, giving a fixed degree of braking, it would be possible to substitute a pot to give variable braking. Such a pot would have to be spring loaded.

The inertia and braking circuits can be operated from signalling circuits so that a train will reduce speed gradually when approaching a signal which is against it and stop at the signal until the latter changes. It will then accelerate slowly and eventually resume its normal speed.

Finally, there is what we have dubbed the "kick in the pants" circuit. One of the most annoying situations is that where a train that will not start when a normal starting voltage is applied to the tracks. At best, it will not start until a lot more voltage is applied, at which point it will start with a violent jerk. At worst it will not start at all, meaning that it has to be pushed with the finger and the whole illusion destroyed.

One obvious cause of this problem is poor contact between the rails and loco wheels, but there appear to be more subtle causes. To quote just one example; a motor will sometimes start in reverse when it will not start in the forward direction.

Our "kick" circuit is designed to minimise this problem by applying full voltage to the track for a fraction of a second; just enough to get the motor armature moving. The voltage then reverts to that selected by the control knob or the inertia circuit.

In practice the idea works extremely well. The loco starts with a slight jerk – the exact amount can be selected – then picks up speed slowly as dictated by the inertia circuit. The effect is not unrealistic, since the jerk is minimal and normally barely sufficient to take up the slack in the rolling stock couplings.

On the other hand, the starting reliability is very high. Even when a loco fails to start the first time, a second kick – easily applied – will often be effective. In extreme cases the situation can sometimes be salvaged by switching to reverse, applying a kick without follow-up power, then switching to forward and trying again. Even this complex procedure is preferable to jabbing the loco with one's finger.

A basic circuit is shown in Figure 1. The first point to note is the filtering network provided by D1 and C1. A filtered supply proved essential for the control circuitry and, since it was not practical, or even desirable, to filter the supply to the rails, we simply filtered what we needed. The diode/capacitor combination is a particularly effective one.

Inertia is provided by C2, in the base circuit of TR1. The two 5.6k resistors (R1, R2) each side of it perform a number of functions. Collectively they serve as a current limiting resistor for the base of TR1. Separately, they control the rate of acceleration and deceleration. R1 controls the rate at which C2 can charge to the voltage set by the moving arm of the throttle pot (R3). R2 controls the rate at which C2 can discharge through the base circuit of TR1.

Their total value also has some bearing on the spread of the throttle movement. If the value is too low TR1 and TR2 may saturate before the throttle is fully advanced, making its operation critical. If too high, the power available to the track willbe limited. The values chosen are a compromise between all these requirements.

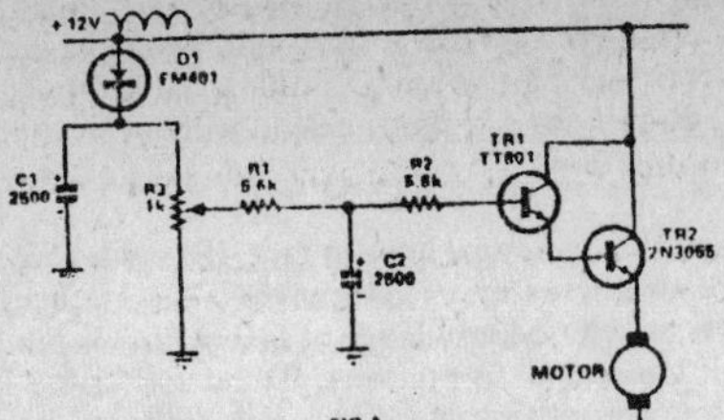

FIG. 1

These two simple circuits are quite practical, and may be built as the first step towards our final circuit.

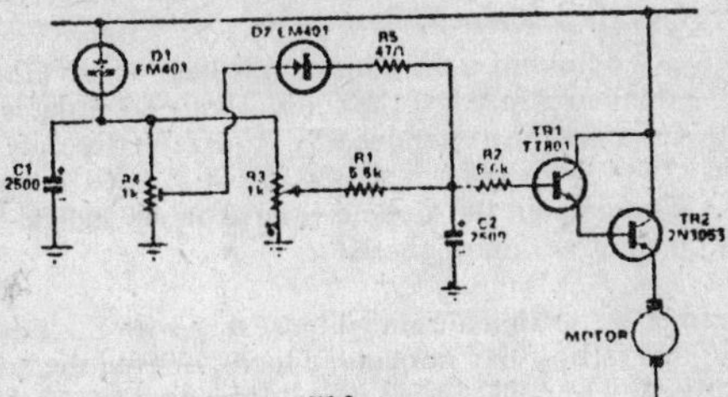

FIG 2

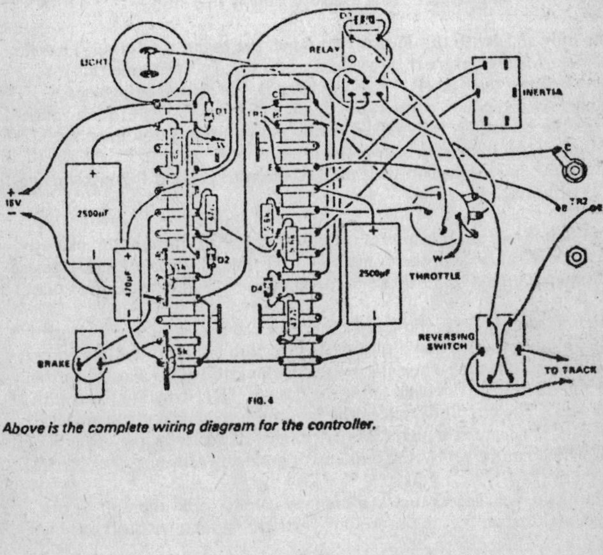

FIG. 4

Above is the complete wiring diagram for the controller.

While the amount of inertia could be varied by varying these resistors, this is not recommended in view of the factors just mentioned. The better approach would be to vary C2 although, in practice, the value we have chosen should suit most situations. In this regard it must be remembered that the final circuit has a brake function which can override the natural deceleration function.

This simple circuit is quite practical and, in fact, the reader may care to build it to this point as a first step. This approach is particularly recommended for those whose knowledge of electronics is minimal. However, the circuit has several limitations. The most serious is that when the throttle is advanced from zero the train hesitates for a time until the voltage increases to the minimum necessary to start it. To overcome this problem it was decided to prevent the inertia capacitor from discharging completely. This is accomplished by adding a 1k tab pot (R4), a diode (D2), and a 47 ohm resistor (R5) (Figure 2). The setting of R4 determines the minimum voltage across C2 and, therefore, the minimum speed of the train. Diode D2 isolates the two circuits so that the voltage determined by R4 is effective only when it is greater than the voltage determined by the throttle (R3). Once the voltage from R3 is greater, the diode is reverse biased and ceases to conduct. R5 is simply a limiting resistor.

In use, R4 can be set so that a train will just stop with the throttle fully retarded, but one setting may not suit all locos, or even the same loco when it is either hot or cold. In practice it is best set for a minimum speed, the brake circuit being used to bring it to a stop.

The only snag with this idea is that when the brake is released the train may start again before it is required to do so. To overcome this we modified the throttle (R3) to a switch pot, the switch being used to open-circuit the base connection of TR1 (Figure 3). With this arrangement the throttle is first moved to the minimum position until the train slows to minimum speed, the brake applied to stop it, and the throttle then switched to the "off" position to prevent it from starting prematurely.

This switch has another useful function. It will stop the train instantly and is very valuable in the event that a possible crash is likely to damage an expensive model. In these circumstances realism takes second place.

The braking system is shown in the final circuit (Figure 3). It consists of a 5k tab pot (R6) and a push-button switch between the base circuit of TR1 and chassis. When the switch is closed C2 is discharged through R6 and all voltage removed from the base of TR1, bringing the train to a half. The circuit overrides the minimum speed circuit involving R4, etc. The rate at which C2 is discharged depends on the value of R6, which can be set by the builder for greatest realism.

Once again, we suggest that readers may care to build the unit to this stage, and test it, before proceeding further. In fact, we built our own

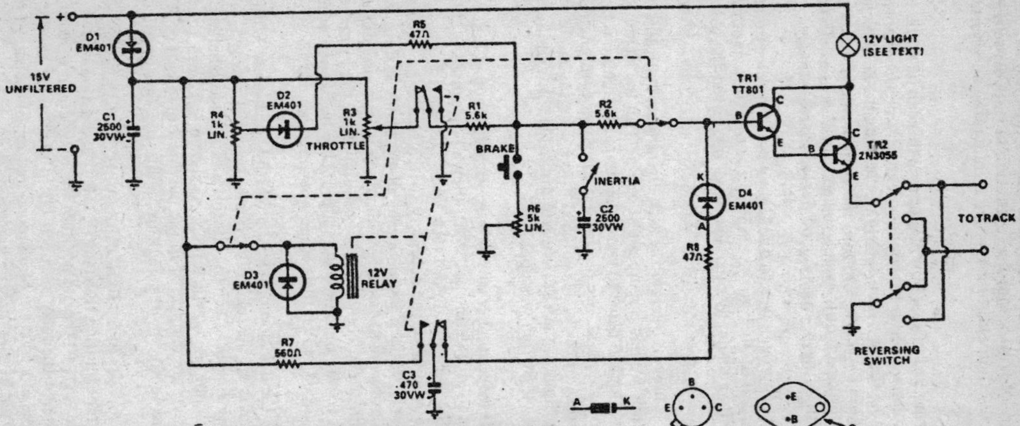

FIG. 3

TRANSISTOR TRAIN CONTROLLER
WITH INERTIA

The final schematic diagram of the controller incorporates the "kick in the pants" starting circuit and improved braking facilities.

unit to this stage and experimented with it for some time before evolving the next feature.

This is the "kick in the pants" circuit previously mentioned. The idea was born largely of frustration; the frustration of both ourselves and most other model train enthusiasts when a train refuses to start on cue.

Having been reminded of just how frustrating this can be while testing the new controller we began to wonder whether something could be done about it. At first we tried simply discharging the largest capacitor we could find directly across the rails. While impractical, it did prove that a kick of this kind would start a stalled loco almost as effectively as jabbing it with a finger – and a good deal more realistically.

A more practical approach was to discharge a much smaller capacitor into the base circuit of TR1 at the moment the throttle was advanced from zero. Again, a manually operated mock up proved the worth of the idea.

The problem was how to operate the kick circuit automatically from the throttle. While we had a spare set of contacts on the switch pot, and this seemed a logical place to start, the simple switching action it provided was not adequate. We needed a changeover combination so that the kick capacitor could be switched into a charging mode while the train was stationary, then switched to discharge through the base circuit when the pot switch was operated.

While we considered a number of schemes, including added extra switch functions to an existing pot shaft, they all presented problems of one kind or another. Finally we took what was really the easy way out, and added a relay with the necessary switch functions. While it might be possible to devise a more elegant all-electronic system, we reasoned that relays were something with which most train enthusiasts were readily familiar and could probably supply from their junkbox. Even if they had to buy one, they would probably be happier with it than a complex circuit which they could not understand.

As it turned out, the decision provided a worthwhile bonus. With an extra set of contacts on the relay – normally provided anway – it becomes relatively simple to use the control unit with a signalling system while retaining the inertia effect. As shown in Figure 3 the extra set of contacts are connected in the moving arm circuit of R3. We will have more to say about using this facility in a future article.

The kick circuit consists of a 470uF capacitor (C3), the relay changeover set, two current limiting resistors (R7, R8) and a diode (D4). When the throttle is set to the off position the relay coil is disconnected by the pot switch and the relay connects C3 to the filtered supply rail via R7. C3 charges in a fraction of a second and remains charged until the relay is activated by the throttle being advanced.

When this happens, C3 is connected directly to the base circuit of TR1, and, for a very brief period until it discharges, switches TR1 and TR2 into full conduction. As a result, full voltage is applied to the loco for long enough to start the motor moving, but not long enough to produce any significant high speed movement of the train. As C3 discharges, the minimum speed circuit takes over, followed by the inertia circuit.

Overload protection is by a filament lamp, functioning as a non-linear resistor in the collector circuit of the Darlington pair. During normal running conditions, the current in the lamp is a small fraction of its normal rated current, and the lamp resistance and the voltage across it are very low. This means that the voltage being applied to the loco is not measurably reduced.

During overload conditions, such as when the loco is derailed and short circuits the rails, the current through the transistor and, the lamp increases. The resistance of the lamp increases by a factor of about eight, so the voltage across it increases to a value close to the normal rated voltage. This limits the voltage across the transistor to a safe value. The light output from the lamp can be used as a visible overload warning by placing it beneath a red bezel.

The wattage rating of the lamp depends on the power consumed by the train. It must be large enough so that under normal running conditions the filament does not heat to any noticeable extent – this can be checked by observing the filament, which should not glow at all. On the other hand, the wattage should not be so high that the maximum permissible ratings of TR2 are exceeded. In practice, with the largest lamp likely to be readily available, these ratings are not even approached.

We fitted a 12V 50/40W twin filament car headlight lamp, with both filaments in parallel. This should be large enough for almost all likely situations, yet provides adequate protection for TR2. The only real objection to this arrangement is the cost of the lamp although it should be appreciated that, in this role, it should have an indefinite life.

A possible compromise is to use a discarded lamp in which one filament has been burnt out. This may restrict performance slightly for larger locos, or when double heading, but not seriously. The next compromise is a 21/6W stop tail light, with both filaments used to make a 27W lamp. This would be suitable for use only with small locos, but, if two such lamps are used in parallel, making a 54W unit, performance will be at least as good as from a single filament headlight lamp. Since these lamps cost little, they are a much better proposition financially.

In constructing the unit, we elected to fit all the components onto a single front panel. The idea was that the panel could be fitted easily into an existing control layout, or it could be used as the lid or top panel of an available box. We used the latter idea, and dimensioned the panel to suit a standard sloping panel instrument case (approx. 6 in x 6in x 6in) such as we used for our previous controllers.

This is a suggested layout only, and constructors may wish to vary it. We found that as the design evolved, there was not enough space to mount all the components, so a right-angle bracket was used. This was attached to the panel by the overload bezel and the brake switch, thereby avoiding unsightly screws on the front panel. To this panel was fixed the relay and the two tag strips.

The wiring and layout of components on the tag strips is not critical. Our layout is shown in Figure 4. The preset pots should be positioned so that they are readily accessible for adjustments. We found it an advantage to use coloured wires, as this makes the wiring easier to trace when it is being checked.

We mounted the light bulb using heavy gauge tinned copper wire, which provides both electrical connection and mechanical support.

When the unit is completed and checked, it can be connected to a suitable supply, being careful to ensure the right polarity. The output can then be connected to a track, preferably a circular one as continuous running facilitates testing, and a train placed on the rails. The inertia is switched off, and R3 set to the minimum speed position. (Not the off position.)

The minimum speed is then set by the preset pot, R4. It is set so that the train will run steadily at a slow speed, but so that when the train is stopped, say by momentarily touching it with a finger, it will remain stopped.

The "kick in the pants" effect can now be tried, by switching R3 from minimum to off and back again. The train should stop when switched to off, and start with a small jerk when switched to minimum speed. If this jerk is too violent, C3 can be reduced in value. If the train does not start reliably, the value can be increased.

It should now be possible to vary the speed of the train smoothly from minimum to maximum by R3. When the train is at full speed, the track can be shorted out. The train should stop, and the overload indication should glow. When the short is removed, the train should move again.

With the train stationary, and the speed control in the off position, the inertia can be switched in. Allow a few seconds for the inertia capacitor to charge to the voltage set by the minimum speed pot, and then advance the speed control to the full speed position. The train should start immediately with a small jerk (due to the "kick in the pants"), and then accelerate slowly to maximum speed, which should be reached in about 15 seconds. When the speed is reduced to the minimum setting, the train should slow to its minimum speed in about the same time. It should not stop.

With the train running at minimum speed, depress the brake. This should stop the train in a short distance. This distance is set by R6 and can be varied to suit individual requirements.

Alternatively, the brake can be adjusted with reference to the distance required to stop from full speed. With the train going at full speed, and the inertia switched in, set the speed control to minimum, and apply the brake. The train should stop in a distance that is realistic in comparison to the distance taken to coast from maximum to minimum speed.

The power supply required to run the train and controller should be capable of supplying a nominal 12V DC, with a current rating of from 2 to 5 amps. Best performance will be obtained if the supply is not smoothed. We used the "Train Power Supply", set on the 15V tap. The normal commercial "train transformer/rectifier" should be quite suitable, since the majority of these units consist of a 15V transformer and associated rectifier.

It should be remembered that best performance will only be obtained from the controller if the rails are kept clean and free from dirt, to minimise contact resistance. Rail joint resistance is also a cause of poor performance. This can be minimised by connecting the power to the rails at several points, normally spaced about 1 metre apart.

List of Component Parts

SEMICONDUCTORS

4 EM401 silicon diodes, or RS276-1139.
1 TT801 NPN silicon transistor RS276-2018.
1 2N3055 NPN silicon power transistor, with insulating mounting kit (RS276-2023, BD130, BDY39).

SWITCHES

1 Single pole, to match. 1 2-pole, double throw.
1 Spring loaded single pole, normally open.

CAPACITORS

1 470uF 25VW electrolytic. 2 2500uF 25VW electrolytic.

RESISTORS (¼W unless specified otherwise)

2 47 ohms.
1 560 ohms.
2 5.6k
1 1k preset pot, linear.
1 1k potentiometer with switch (double pole), linear.
1 5k preset pot, linear.

MISCELLANEOUS

1 Panel, see text.
1 12V 40/50 watt double filament headlight lamp (see text).
1 12V miniature relay, with two changeover sets.
1 Knob.
1 Bezel, red.
2 13 bit tag strips.
Hookup wire, solder, screws, nuts, 18 gauge tinned copper wire, terminals if required, case if required (see text).

SIGNAL SYSTEM FOR TRAIN CONTROLLER

Here is a circuit to couple your model train controller to a signalling system. The full inertia characteristics are retained, producing a most realistic effect as the train decelerates and stops, then starts and accelerates to its original speed – all automatically in response to the signal.

We presented a Model Train Control with Simulated Inertia. This was designed to overcome the lack of inertia in a model train, in both the starting and the stopping modes. We also incorporated a "kick in the pants" circuit to help overcome the problem of non-starting trains.

In this article, we are presenting an add-on circuit to incorporate automatic control of the train in conjunction with a signal. What we had in mind was to provide a switch to control a signal, and to have the controller interlocked with the switch, so that the train would obey the signal.

In fact, we had this thought in mind when we originally designed the controller. It was the reason we used the second change-over contact set on the "kick in the pants" relay to simulate setting the throttle to minimum speed. In Figure 1, RL1 disconnects R1 from the wiper of R3 and discharges C2 (slowly) through it.

This means that a switch placed in series with the relay coil can be used to stop and start the train, with full inertia effects, without disturbing the main throttle setting.

Consider the case where the train is travelling at some arbitrary speed. The throttle will be at the appropriate setting, and the relay in the energised position. If the relay coil circuit is opened, the lower change-over set (Fig. 1) will connect C3 to the positive supply rail via R7, charging it so that the kick in the pants may be applied when the relay is energised again.

The upper change-over set will effectively set the throttle to the minimum speed position, so that the train will slow and eventually reach the minimum speed as set on R4. At this stage we may apply the brake, and bring the train to a stop.

When the relay is re-energised, the throttle is returned to the original setting and, at the same time, the kick in the pants is applied. This starts the train, which slowly accelerates to its original speed. Thus the obvious place to interconnect our signalling circuit is in the coil circuit of relay RL1.

To stop the train completely after it has decelerated we may apply the brake manually but, fairly obviously, it would be much better if this also was an automatic function. This is not hard to arrange and is simply a second braking circuit in parallel with the first. The two circuits remain independent and we can make the automatic brake fiercer or softer as required.

The next problem is to sense the position of the train. Perhaps the simplest solution, and one that we have used in the past, is to use reed switches in the tracks in conjunction with permanent magnets underneath the trains. As the train passes over the reed, the contacts close for a short period. This is sufficient to indicate the presence of the train.

Since the reed contacts close for only a short time while the magnet is passing, it is necessary to provide some sort of latching device to "remember" the passage of the train. This can be done quite simply, in either of two ways. We have used both.

The first method is to have the reed operate an electromagnetic relay in the latching mode. The reed switch is connected in series with the coil of the relay, so that the relay is energised when the reed is closed. A normally open contact set (one that closes when the relay is energised) is connected in parallel with the reed, so that the relay will "lock on". Other contact sets are used to perform the required switching functions. The relay may be unlatched by momentarily opening the coil circuit.

The second method is based on the different pull-in and drop-out ampere turns of a reed switch. It is a characteristic of a reed insert that a larger value of ampere turns is required to operate the reed initially than is needed to hold it closed. With the reed we used, 30 ampere turns are required to close the contacts, but the contacts do not open till the field is reduced to 12 ampere turns.

This means that the reed will not operate in a 20 ampere turn field but if this field is reinforced by an external permanent magnet so that more than 30 ampere turns are present, the reed will close. When the external magnet is removed, the reed will stay closed. The reed is thus latched on, and the required switching functions may be performed by the closed contacts. The 20 ampere turn field is most easily provided by placing the reed inside a coil.

The basic circuit for stopping and starting the train is shown in Figure 2. Connections between the controller and the auxiliary circuitry are via 5 pin DIN plug and socket. The socket was wired into the controller as shown in Figure 1.

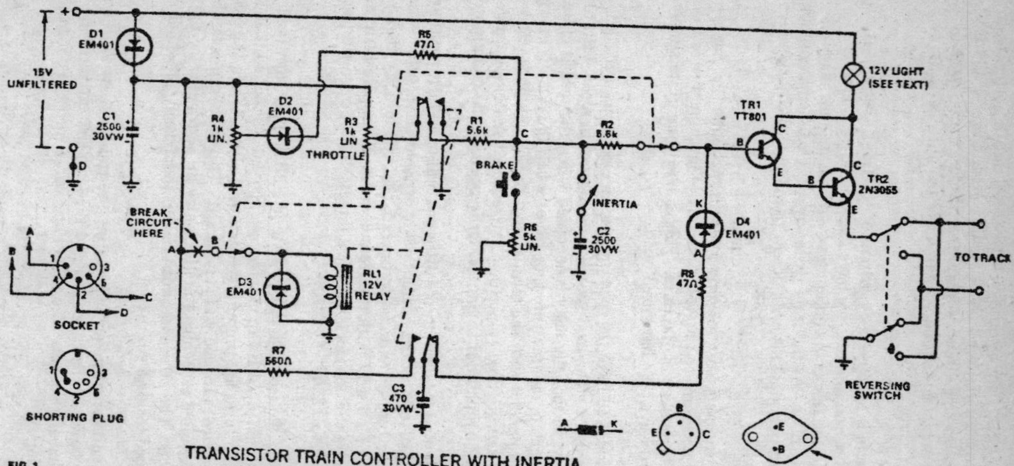

FIG. 1

TRANSISTOR TRAIN CONTROLLER WITH INERTIA
(MODIFIED FOR USE WITH SIGNALLING UNIT)

This circuit is substantially the same as that published The main change is the addition of a 5-pin plug, wired as indicated by the letters "A", "B", "C", and "D". Note the break in the original circuit between "A" and "B".

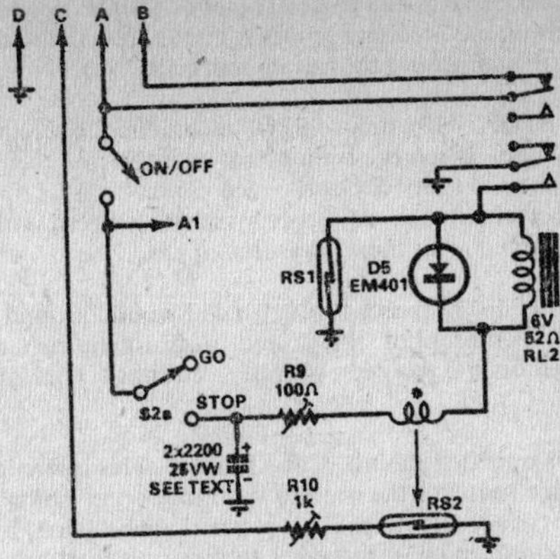

*330T, 36 B & S ENAMELLED COPPER WIRE, WOUND IN TWO LAYERS ON REED, EACH APPROX. 20mm LONG.

SIGNALLING CIRCUIT

FIG. 2

How the sensing reeds and the control relay are connected. RS1 initiates the slowing down action, RS2 operates the brake.

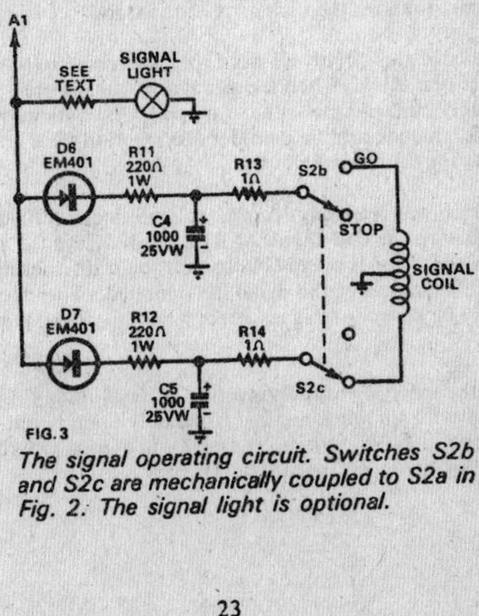

FIG. 3

The signal operating circuit. Switches S2b and S2c are mechanically coupled to S2a in Fig. 2. The signal light is optional.

When the STOP/GO switch (Fig. 2) is set to the GO position, the relay RL2 remains unoperated and points A and B are connected together. This allows the controller to operate normally.

Setting the STOP/GO switch to STOP commences the sequence of events to stop the train. When the train passes over the first reed switch, RS1, the magnet momentarily closes the reed contacts. This operates relay RL2 and a set of contacts on this relay shorts the reed contacts, holding the relay operated after the reed switch opens.

The relay coil current passes through a coil wound around the second reed switch, RS2. A preset pot, R9, is used to adjust the current to hold RS2 operated once it has been operated, but not sufficient, alone, to operate it.

When RL2 is operated, points A and B open. This interrupts the current to RL1, which switches the train to its decelerating mode, as explained earlier. The first reed switch, RS1, is positioned so that, by the time the train is approaching its intended stopping place, it has slowed to its minimum speed.

When the train reaches the second reed, which is placed just before the intended stopping point, the magnet closes the reed, which latches on. This connects R10 across C2 (Fig. 1) and brakes the train to a standstill. R10 is adjustable, to allow for variations between trains.

To start the train again, the STOP/GO switch is set to the GO position. This opens the circuit to the coil of RL2 which, in turn, supplies power to RL1, and initiates the kick start of the train.

When we tried this circuit we were very impressed with the realism of it, except for one thing. When the signal arm was lowered the train started immediately, almost before the arm had finished its movement. Since, in real life, there would be considerable reaction time on the part of the driver, we sought to simulate this.

The solution was relatively simple. A large electrolytic capacitor is wired in parallel with the coil circuit of RL2. This delays the relay dropping out when the signal is set to GO, so that the train does not start for about one second after the signal has operated. The effect is very realistic. The capacitor has no effect when the signal is switched to STOP.

Much of the experimentation involved in developing this system was concerned with finding a method to reliably operate our-signal, using a simple two position switch. Here we will digress to explain the operation of the signal unit itself.

The signal consists of the usual moving arm type indicator, operated by a lever inside the mast. At the bottom of the mast is a centre-tapped coil. Inside the coil is an iron core, attached to the operating lever. When the upper section of the coil is energised, the core is drawn upwards, and the lever moves the signal arm to the GO position. When the lower section of the coil is energised, the core moves downwards, and the signal arm moves to STOP.

The mechanical parts of the system are purposely made so that a small amount of friction is present. This means that, once the signal is moved to either of its positions, there is no need to keep the coil energised. In other words, the signal is essentially an impulse device, and does not require power to be applied continuously. In fact, continuous power would be detrimental to the coil.

Our problem was to relate these requirements with those of Figure 2, assuming that contacts S2a were operated by the same switch mechanism as that used to operate the signal. Whereas the stopping and starting circuitry requires the switch to remain in one or other of its two positions, the signal requires only a fleeting contact or other pulse generating circuit.

The circuit we evolved does this quite well. It uses two poles of a three pole two position switch, connected as shown in Figure 3. With the switch in the position shown, C4 is charged via D6 and R11. The bottom section of the coil has a small current passing through it, as determined by R12 and R14 in series. This is not sufficient to operate the signal, which would be in the STOP position. C5 is charged only to a small potential, as the ratio of R12 to R14 is large.

When the switch is moved to GO, C4 discharged via R13 through the top of the coil. This moves the signal to the GO position. At the same time, C5 is charged via D7 and R12, ready to return the signal to the STOP position. Diodes D6 and D7 isolate any pulses which appear, by transformer action, at the opposite end of the coil.

Resistors R11 and R12 are necessary to isolate the supply rail (A) from the signal coils. They restrict the charging rate of C4 and C5 but, with the values shown, maximum operation rate of the signal is about once per second; plenty fast enough for any likely situation. Resistors R13 and R14 limit the maximum current through the switch contacts during the discharge of C4 and C5. They also prolong the pulse in the coil, giving a smoother action.

We combined the stopping circuitry and the signal control circuitry in the one unit. We used a three pole two position switch; two poles to control the signal, and the remaining pole to actuate the stopping circuit.

Our signal was fitted with a small lamp, which we wired as shown. We have not given a value for the associated resistor, as this will depend on the rating of the lamp. We also included an ON/OFF switch (Fig. 2) to disconnect the signalling circuitry. This switch controls the power to the signal, the signal light and to RL2.

The next point is the relative positions of the signal and the associated reed switches. We placed the signal just before a set of points. The first reed switch, RS1 was placed about 4m before the signal, while the second reed switch, RS2 was placed about 0.25m before the signal. The reeds were placed at right-angles to the sleepers. As RS2 has a coil wound on it, it was necessary to cut away some sleepers so that it could be lowered, otherwise the train would not clear it. This also helps conceal the reed.

We positioned the magnet under the tender of the train, as this was the most convenient location. We used a cardboard spacer, and glued it in position so that the magnet was just above axle level.

The coil on RS2 was made from approximately 330 turns of 36B&S enamelled copper wire, wound in two layers approximately 20mm long. We used the reed itself as the former, and held the wire in place with tape. We did not count the turns, but wound on as many as would fit on the two layers. The wire size is not critical, but should be as small as possible, as this makes the coil easier to conceal.

Once construction is complete, testing can commence. Place the ON/OFF switch in the OFF position, and check that the controller operates normally. If it fails to do so, the most likely cause is an open circuit in the coil circuit of relay RL1. Check that there is a continuous circuit from point A in Figure 1 through the connecting cable and the contacts of RL2 back to point B.

Remove the train from the rails. Place the ON/OFF switch in the ON position, and operate the STOP/GO switch. Check that the signal arm moves, and that it is in phase with the switch. If not, reverse the connections to the ends of the signal coil.

Next, adjust R9 for correct operation of reed switch RS2. Switch to the STOP position, and push the tender by hand over the first reed switch, RS1. RL2 should operate, and remain energised after the tender is removed. If it fails to do so, check that R9 is in the minimum resistance position.

Disconnect the wire from the end of RS2 nearest to R10, so that RS2 is isolated from any voltages, and connect an ohm-meter or similar device across it. With R9 in the minimum resistance position, the contacts should close. If not, then either more turns will have to be provided, on the coil, or a lower resistance coil used with RL2.

Slowly increase the value of R9 until the contacts open, and then decrease the value by a small amount. Push the tender over the reed RS2 and associated coil, and check that the contacts close and remain closed. If they fail to do so, reverse the tender and try again. This is necessary because the magnet in the tender will assist the coil in some positions and hinder it in others, as it moves over the coil. The polarities must be arranged so that the final influence of the magnet is to assist the coil.

If, when the tender (and hence the magnet) is reversed, the contacts stay closed, then either the magnet or the coil will have to be reversed.

If reversing the tender does not close the contacts, decrease the value of R9 and try again. This process will have to be continued until the reed will reliably latch-in when the tender is pushed over the reed.

Set the brake resistor, R10, to minimum resistance and set the signal to GO. Place a train on the rails, and allow it to run at the usual maximum speed. At a time when the train is just about to cross the first reed, change the signal to STOP. When it passes the reed, the relay RL2 should be heard to operate. The train should then commence to slow down.

Provided there is sufficient track, by the time the train reaches the second reed, RS2, it should be travelling at it's minimum speed. When RS2 latches on, the train should stop immediately. If it fails to do so, R9 will have to be re-adjusted. Once it has been verified that RS2 is latching on, R10 may be adjusted to give a more realistic braking performance.

We found that we could adjust R10 so that the train would stop right at the signal. The final effect of the complete circuit was very impressive. If the reeds are concealed, the train appears to stop in quite a realistic way, giving the impression that it really is controlled by a driver in the cab! The delay on starting produced by the capacitor in parallel with RL2 adds to this impression.

SUPPRESSING RF INTERFERENCE FROM MODEL RAILWAYS

Do your neightbours complain loudly about buzzing noises in their radios when you operate your model trains? Does your TV picture suffer from little black lines when your children "play trains"? If so, then you can probably obtain relief by fitting RFI suppression components to your model railways.

Radio frequency interference, or RFI for short, is not usually of concern in the average household, as items like toasters and electric jugs do not usually suffer any detrimental effects when even large amount of RFI are present. However, as more and more electronic equipment, in the form of TVs, calculators and other similar products, is brought into the home environment, effects can be observed due to RFI.

One of the most commonly seen interferences occurs when an electric model railway is operated in close proximity to a television set. The usual effect is for the picture to be covered with little black lines, which detract greatly from the pleasure (or otherwise) of watching TV.

These effects, and other similar ones, can be eliminated by applying modern RFI suppression techniques. Before discussing the suppression circuits required, we will digress and give a short summary of the way in which RFI can reach a susceptible device, and the relevant international and Australian standards.

In general, there are two ways in which RFI can reach the radio, TV set, or other electrical equipment. At frequencies below 30MHz, the interference is primarily conducted along the mains wiring of the radiating device, and hence into the mains system itself. It is then coupled either inductively or capacitively into the electronics of the susceptible equipment.

The second way in which RFI can be propagated is by direct radiation. This occurs principally for frequencies above 30MHz.

To be effective, any suppression system or technique must eliminate or reduce both types of RFI to acceptable levels, as specified in the relevant standards.

The international standards for radiated and conducted interference levels are asllaid down by the Comite International Special des Perturbations Radio electriques (CISPR).

AS 1WG covers the electromagnetic spectrum in two ranges, from 150kHz to 30MHz, and from 30MHz to 300MHz. Limits are given for the allowable RFI in terms of terminal voltage for the lower band, and in terms of interference power and field strength for the upper band.

Table 1 gives the allowable limits for conducted RFI, measured in the approved manner at the mains terminals of the offending equipment. Where there is a frequency duplication at the boundary of the ranges, the lower limit applies.

For frequencies between 30MHz and 300MHz, the limits are specified in terms of both power and noise field strength, to allow for both power and field strength measurements. The appropriate limits are given in Table 2.

High frequency interference with radio and TV reception occurs wherever commutation processes generate voltage or current transients. The main RFI sources in a model railway are the locomotive motor itself and the varying current transmission paths between rails and wheels and contact springs and wheels. The high frequency interference signals so generated propagate along the rails and supply leads, and reach the susceptible device by a combination of inductive and capacitive coupling, and by direct radiation.

TABLE 1
LIMITS OF TERMINAL VOLTAGE IN THE RANGE 0.15 TO 30 MHz

Frequency	Limits	
MHz	Hand held portable tools uV	Other appliances and equipment uV
0.15-0.2	3000	2000
0.2-0.5	2000	2000
0.5-30	1000	1000

TABLE 2
LIMITS OF INTERFERENCE POWER AND NOISE FIELD STRENGTH

Frequency	Limits	
MHz	Interference power dB(pW)	Noise field strength dB(uV/m)
30-300	45 rising linearly with frequency to 55	25 rising linearly with frequency to 35

TABLE 3
VALUES OF SUPPRESSION COMPONENTS

Gauge	HO	N
Motor	3uH VHF choke 0.047uF ceramic cap.	2.2 uH VHF choke 0.047uF ceramic cap.
Track supply point or transformer	0.68uF ceramic cap.	0.68uF ceramic cap.

In the frequency range of 150kHz to 30MHz, the interference is primarily conducted. Interference suppression is relatively easy in this band, particularly in smaller sized installations. It generally suffices to connect a 0.1 to 1uF capacitor permanently across the low voltage leads, either at the tracks or at the transformer. This will reduce the conducted interference to below the recommended level. Figure 1 shows interference voltage curves for a small model railway before and after suppression.

Suppression in the 30 to 300MHz band is not so simple. A single suppression capacitor alone is almost entirely ineffective because the interference from the rail or leads and the locomotive itself is directly radiated. To be effective, the interference fields must be suppressed as close as possible to their origin.

The RFI from the motor is not difficult to suppress when the necessary suppression components such as VHF chokes and capacitors are located close to the motor connections.

Suppressing the current transmission paths is more problematic, however, as the rail/wheels interference source continually changes its position as the train moves. Successful suppression must be effective over the entire installation.

Each interruption in current flow at the current collectors, wheels and rails, for instance due to dirt, causes high frequency voltage transients in accordance with $u = L.di/dt$, where L is the rail inductance. Current interruptions cannot be avoided, but the inductive reactance between the rails can be compensated at high frequencies by discrete capacitors, placed across the rails at short intervals.

This method is complicated however, and the capacitor leads themselves are also inductive, making the compensation inadequate.

A uniformly distributed capacitance between the rails as shown in Figure 2 is most effective. The metal layers with dielectric between them are embedded in the ties and form a capacitor distributed uniformly along the rails. The interference suppression obtained with this type of rail is good, as shown in Figure 3. Without suppression, the interference exceeds the recommended standard by 10 times or more.

Interference suppression on the motor with two VHF chokes and a capacitor and suppression at the track supply terminals primarily reduces the interference field strength at low frequencies in the 30 to 100MHz range. Only the use of the suppression rail assures compliance with the recommended limit, however.

Unfortunately, this type of track is not available in Australia, so it will probably not be possible to obtain this degree of suppression. However there is no doubt room for experiment here, with multiple bypass capacitors along the track, and also perhaps home-made distributed capacitors. You could try making such capacitors using tinplate sheet, with thin PVC or mylar as the dielectric, and mounting the assembly underneath the track.

Figure 4 gives a diagrammatic view of the basic suppression components required for a typical model railway. The values required are shown in Table 3. Note that only VHF chokes and low impedance capacitors should be used.

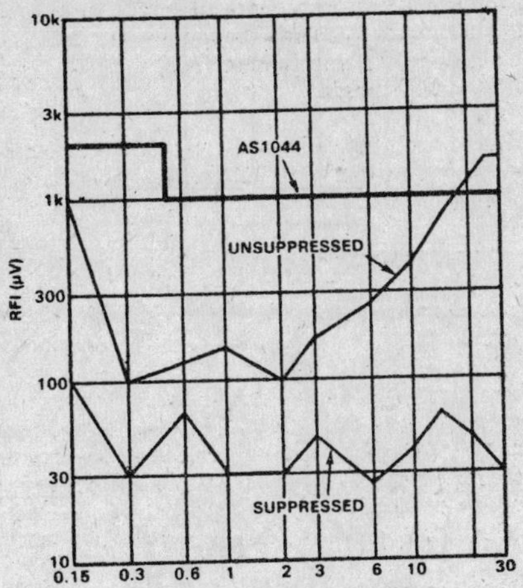

FIG. 1

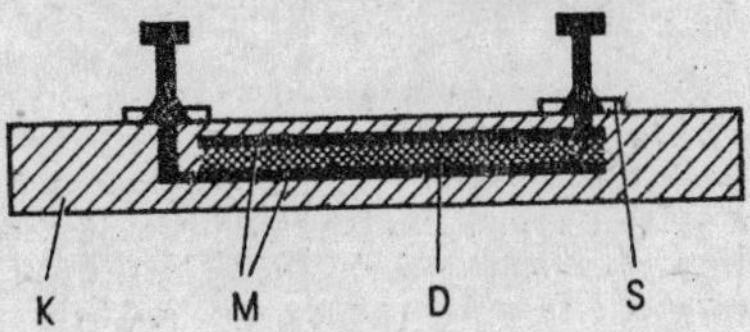

M	Metal layers
D	Dielectric of the suppression capacitor
K	Insulating tie

Fig. 2: Cross-section of the type of track having an integral distributed suppression capacitor.

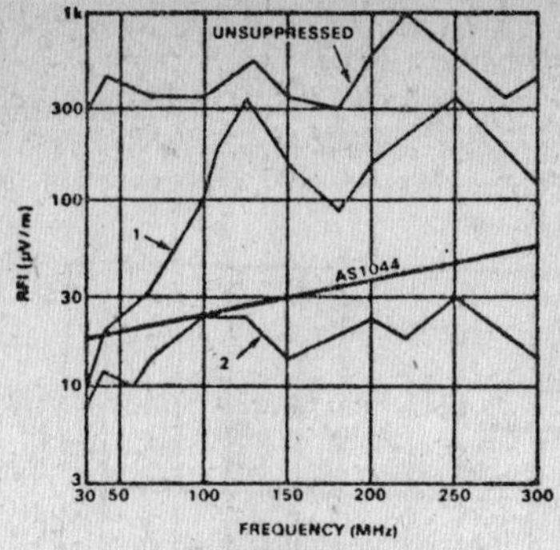

1 = MOTOR AND RAIL SUPPLY SECTION SUPPRESSED NORMAL RAILS

2 = MOTOR AND RAIL SUPPLY SECTION SUPPRESSED SUPPRESSED RAILS

FIG. 3

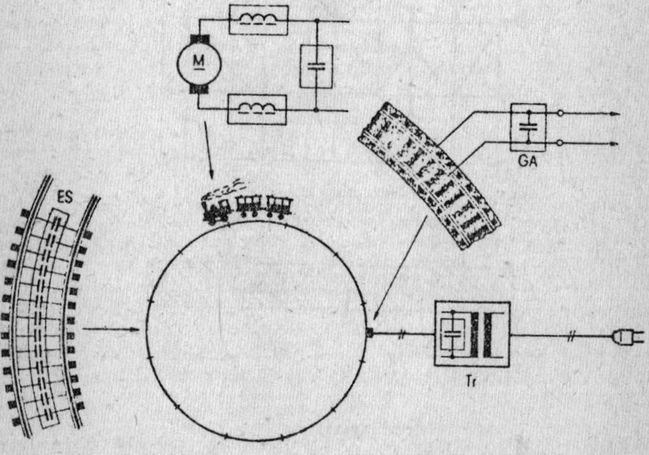

Tr	Transformer	ES	Suppression rail with distributed capacitance, equivalent circuit
GA	Track supply section, suppressed		
M	Motor with suppression components		

Fig. 4 (above): The basic RFI suppression techniques suggested for model train layouts. Component values are given in table 3.

LIGHTING FOR MODEL TRAINS

N.B. Certain Australian manufacturers part numbers may be quoted in this book for items of hardware such as bulbs, nickel cadmium cells, switches and magnets etc. These part numbers may not be applicable in the U.K. Please be guided by your dealer as to suitable alternatives.

This article considers provision of continuous constant level lighting for model trains, suggests a relatively simple method by which this may be achieved, and discusses the availability and characteristics of suitable components.

Model railway enthusiasts who aspire to maximum realism in their layouts have always been conscious of the problem of train lighting. While it is easy enough to connect a lighting system across the track and light the train while ever it is moving, the lights will dim, then fail completely, as the train slows and stops. One attempt to overcome this employed a high frequency oscillator to generate lighting power, isolating this from the loco power and loco motor by frequency selective circuits within the loco and between the power supplies.

While an ingenious idea, most enthusiasts seem hesitant to adopt it. Among the reasons given are, (a) it is too complex unless one is fairly well experienced in electronic circuitry, (b) the need to use modified locos means the system cannot accept "visiting" locos, brought by fellow enthusiasts to add to a night's fun, and (3) it can adversely affect certain signal systems.

Although we had always appreciated the problem, the incentive to do something about it was given impetus by a letter from a reader. He had built one of our control units and, while very happy with the performance in regard to train control, pointed out that they made the lighting problem even worse. Due to the pulsed nature of the power provided by these units, the lights not only dimmed at low speed, but flickered badly. Could we suggest how this could be overcome, perhaps by using a capacitor in some manner?

Thus encouraged, we decided to try an idea which we had been considering for some time; an idea which would not only overcome the flickering, but give constant lighting at all speeds, or stationary, and without the complication and other disadvantages of the high frequency system. This article is the result.

Basically, the idea is to equip a train or loco with its own rechargeable battery, with circuitry to enable the battery to be charged from the track while the train is running at normal speeds, but preventing it from discharging back into the power supply system when the train is reversing, running slow or stationary. During these periods, the battery carries the full lighting load.

A few years ago the idea of equipping a model train, particularly in the HO size, with a rechargeable battery would have been out of the question, due to the problems of size, mess, etc. Now, with the advent of the nickel-cadmium sealed rechargeable cell, the situation is quite different. These cells are available in a wide range of sizes, from tiny "button" types used in miniature hearing aids, to standard "D" size cylindrical types, or even larger. They are a hardy cell, not adversely affected by long periods of idleness in a discharged condition, which makes them ideally suited to the intermittent demands of hobbyists generally.

Electrically, the system is delightfully simple, as illustrated in the accompanying circuit. A small power diode functions as an electronic "cut-out" to prevent the battery from discharging when the voltage is reduced below the level at which it can charge the battery. The other major component is a series resistor to control the charge rate. This latter is selected on the basis that the power from the track will carry the full lighting load at normal running speeds, as well as charge the battery at its maximum permissible rate. This, ideally, should offset the periods of discharge when the train is running slow, stationary, or reversing.

In putting such an idea into practice, a number of decisions have to be made. A major one concerns the voltage at which the lighting system will operate and this, in turn, will be governed to a large extent by the types of lamps available, or dictated by the physical characteristics of the model. In this regard we can only make general comment, leaving it to the individual modeller to make up his own mind.

However, we suggest 1.25 and 6 volts as being the two most suitable values. with a preference for 6 volts for all but the smaller installations. The lower voltage has the advantage that only a single cell is required and, for an equal number of watt-hours, will probably be cheaper than a multiple cell system. On the other hand, the higher current required at the lower voltage can present practical problems where more than one or two lamps are involved.

One of these concerns the series resistor used to control the charging rate. To lignt a complete train at 1.25 volts would require up to 1.5 amps and, while there is a small enough cell available which can deliver this order of current, it must be remembered that we plan to supply this power from the track most of the time. Thus this current must pass through the series resistor, and the power which the latter will dissipate presents an awkward problem, particularly if the heat is being generated inside a plastic model. Some improvement results from slinging the resistor underneath the waggon, but the problem is still serious at the levels contemplated above.

Another problem aggravated by the higher current is switching. An "on-off" switch for the lamps can take many forms, depending on the ingenuity of the modeller, but there are a number of advantages to be had from using dry reed switches, operated by magnets in the track. The current rating of these is limited to about 250mA (DC) for most types and, while it may be permissible to exceed this a little in applications like this, a practical limit is reached where the reeds tend to stick in the closed position, due to a tiny weld between them. They can be freed by tapping, but the condition is obviously undesirable.

For these reasons we suggest that one-cell operation be limited to single units of rolling stock, locos, brake vans, etc., which are best kept self-contained, but which require no more than about 200mA lighting current overall. For complete train lighting, the 6V system becomes virtually essential.

Which brings us to the next question: whether each individual item of rolling stock should be a self-contained unit, or whether a whole train can be treated as a unit, operating from a single battery.

Treating each item as a separate unit has the advantage that it permits maximum flexibility in making up trains, and there is no need for interconnecting cables between units. It has the disadvantage of higher cost, since it is necessary to provide a battery, diode, resistor, and "on-off" switch for each item. It may also transpire that some individual units may not have room for all these, meaning that it must derive its supply from an adjacent unit or not be lighted.

The individual unit approach would seem best suited to general purpose or goods locos, goods brake vans, and similar items which by their nature, are likely to be used in conjunction with a wide variety of other units. Other items of goods rolling stock do not normally require to be lighted.

Treating a whole train as one unit has the advantage of lower cost and, generally, some simplification. It has the disadvantage that the train cannot be broken up without rendering the lighting system inoperative. This system appears to be best suited to complete passenger trains which have a recognisable identity and are, therefore, normally left in this form.

The next point for consideration is the type of lamps required or likely to be available. Lamps used for general lighting, as in carriages, are not critical, within reason, in regard to shape or size. The most important factors are their voltage and current ratings. Those to be used externally, however, need to conform to certain shapes and sizes – allowing for the innumerable tricks which modellers employ to hide or disguise oversize components – as well as voltage and current requirements.

While the use of voltages lower than the traditional 12 or 14 will rule out the range of lamps normally available for model lighting, there is quite a wide range of lamps available if one is prepared to do a little searching. These range from conventional torch, cycle, and dial lamps, through toy lamps, to medical lamps. The latter may not be so easy to come by, and are usually quite expensive, but include some types which are ideally suited to certain applications. Assuming that, by careful handling and under running, they will have a long life, the cost may be justified.

For simple carriage illumination there are several 6V types available. The smallest (electrically) is a cycle generator tail light globe, rated at between 40 and 50mA. Next is a panel indicator lamp rated at 120mA. This is smaller physically than the previous type and uses a type E5.5 base. This is about half the diameter of the standard torch or dial lamp base, which is the E10.

(These figures would appear to refer to the nominal diameter in millimetres.) After this we come to the standard torch and dial lamps, rated at around 250 tp 300mA.

The cycle generator lamp is particularly attractive because it permits using several small lamps in one carriage, rather than one big one. This is important where carriages are complete with interior fittings, often involving partitions between sections of the carriage. In any case, several small lamps will provide more even illumination over a long carriage than one large one. Assuming, for the moment, that we can supply about 900mA from a typical battery, we could light 18 50mA lamps, or between three and four per carriage in a five-car train.

For single cell operation the most attractive lamp appears to be a standard torch type rated at 1.2 volts, 220mA (Eveready type 1151). This uses an E10 base, but has a smaller diameter globe with a simple lens moulded in the end. Although it can be used for general illumination, the lens produces a nicely focused beam, suitable fro a loco headlight. Assuming that the larger part of the lamp can be hidden in some way, such as behind a smoke box door, it would be ideal for such an application.

A lamp which would seem to be ideal for many model applications is the tiny 1.25V type used in key ring lights. Unfortunately, there does not seem to be any local supply of spares at the moment, although the complete lights are readily available. Should the position improve, they could be used in series sets of five across a 6V supply for multi-compartment carriages.

Of the medical lamps available, one of the smallest and most intriguing we have seen is bronchoscope globe. The globe proper is only 1/16in diameter and less than ¼in long. The base is of slightly larger diameter and about 9/16in long, giving an overall length of just over ¾in. We were unable to determine the exact voltage for which they are designed, but they perform quite well between 1.25 and 2 volts and draw between 75 and 100mA. They would seem to be ideally suited for marker lights on the end of carriages or brake vans, assuming that the length can be accommodated inside the van. If used with a 6V system they could be series connected in a set of three, or two with a 20 ohm resistor.

Before leaving the subject of lamps, it is worth mentioning that, for special applications, it is often worthwhile to reduce the physical size of the lamp by removing the base. Remove the solder from the centre connection, break the black insulating bead and, if possible, cut away a little of the brass base with the point of a pair of diagonal cutters. Soak the lamp in methylated spirit overnight, which should loosen the adhesive, leaving only the other soldered connection at the top of the base to be melted in order to free the base completely.

Assuming suitable lamps have been chosen, the battery requirements can be determined. For 1.2 or 1.5V lamps a single cell will be required and for 6V lamps either four or five cells. Four cells will slightly under-run the lamps, but will help reduce the current drain. This may be important in some cases. Five cells will run the lamps closer to their full brilliance but will increase the current drain and perhaps reduce the life of the lamp.

Nickel-cadmium cells are rated in amp hours (AH) or milliamp hours (mAH) and a suitable type can be selected from this information. The two basic rules to apply concern the maximum charge rate which should not exceed the 10 hour rate, and the maximum discharge rate, which should not exceed the 1 hour rate.

Thus, a cell rated at 1AH should be charged at not more than .1A, the rate at which, in theory, the cell would be fully charged in 10 hours. (In practice, due to losses, the time would be 14 hours.) The same cell should not be discharged at more than 1A, the rate at which, in theory, it would be fully discharged in 1 hour. (Again, due to losses, it will deliver this rate for less than one hour. However, the full AH rate should be achieved at the 10 hour discharge rate.)

Even assuming we get only 45 minutes supply at this rate, it should be more than adequate for any model railway application. In fact, it is unlikely that any one train on display would remain idle for longer than 10 minutes. We can therefore select our cells on the basis of an AH rating equal to the discharge rate in amps. Thus, if a system draws 100mA it can be supplied by 100mAH cells. If larger cells can be provided, so much the better. Note, however, that nickel-cadmium cells cannot be connected in parallel.

When ordering any nickel-cadmium cells for this work, be sure to specify that they be supplied with solder tabs, normally indicated by the "T" in the type number. (The C900, fitted with tabs, is an exception.) It is unwise to rely on pressure contacts in an application like this, and no attempt should be made to solder directly to the battery case.

In the button types, sizes range from just under ½in to just under 2in diameter. The smallest ones are too low in rating to be of much interest, while the largest ones are too large physically for the HO system. The largest size which could be conveniently accommodated (.988in diameter) includes 100mAH (B100T). 150mAH (B150T), and 225mAH (B225T). The latter could be used to supply a single 1.2V, 220mA torch lamp or similar where this was suitable.

On the other hand, we managed to pack four type B100T cells into the tender of a Triang "Princess Royal" loco and use it to power a 6V, 120mA indicator lamp mocked up as a headlight. (Strictly speaking, this loco doesn't carry a headlight, but it was an interesting exercise.) With only four cells the lamp was somewhat under-run, and drew less than 100mA, but performed adequately.

Packing the four cells into the tender wasn't as easy as it looked. The tender was large enough but the makers had gone for extreme realism, providing, in addition to the artificial coal, an imitation water tank beneath it complete with sloping top on which the coal normally rests. We finished up with two cells in the water tank, and two more on top, plus the diode and series resistor. We were then just able to replace the artificial coal.

In the cylindrical cells the C450T (450mAH) and C900 (900mAH) are of most interest for this application, the latter being particularly attractive. Both are 17/32in diameter, the C450T being 1-31/32in long and the C900 3-35/64in long. This length can easily be accommodated in say, the brake section of a combined brake van and coach or, possibly, in a smaller (4 wheel) brake van. Four such cells can be fitted in the width and height of most HO rolling stock, and a fifth with a little wangling. The resultant 6V 900mAH battery should provide enough power for most passenger-train lighting.

We also tried a single CH1.5 (1.5AH) cell housed in the kitchen section (slightly modified!) of a buffet car, with the idea of powering six or seven 1.2V 220mA lamps distributed throughout the train. The idea worked up to a point, but served to demonstrate the dissipation and switching problems already discussed. While an excellent cell, and one which can be accommodated, there is some difficulty in making full use of it in this application.

Summing up, we suggest that 1.5V systems are probably adequate and most economical for single units which require to be self contained. The B225T or C450T cells with appropriate 1.2-1.5V lamps would form the basis for such a system. For a typical five-car passenger train we suggest a 6V system based on four or five C450T or C900 cells and 6V, 50mA cycle generator tail-light lamps.

To reduce interconnecting cables between units to a minimum, we strongly recommend that power for the lighting circuit be picked up from the track by the car carrying the battery, rather than attempting to tap into the loco circuit. Keeping these two circuits separate also simplifies matters if it is found necessary to provide auxiliary charging for the battery. Even in the case of a loco, where the battery is probably housed in the tender, it is better to try to arrange for power pickup via the tender wheels, rather than the driving wheels.

Quite obviously, some means must be provided to switch the lights on and off. If this is to be the manual switch the modeller will probably have to improvise his own, since it is difficult to suggest a commercial switch that could be easily concealed yet readily accessible. We must also consider whether the train will be required to run without lights on occasion and, if so, provide additional switching to disconnect the charging circuit. If this precaution is not observed, the battery would be charged at an excessive rate by a factor of about 10 times.

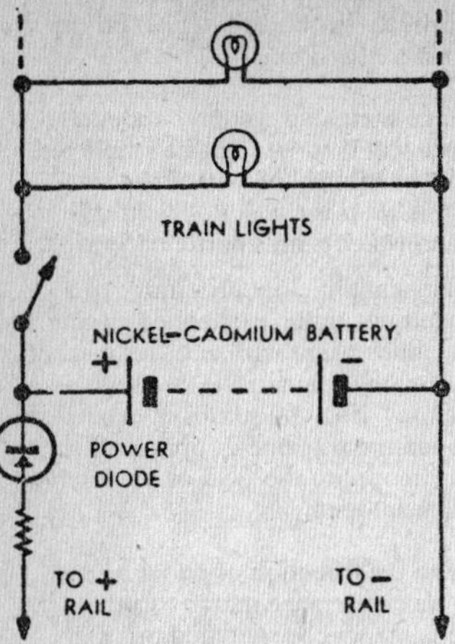

The basic circuit of the train lighting circuit. Methods of determining exact component values will be discussed in detail

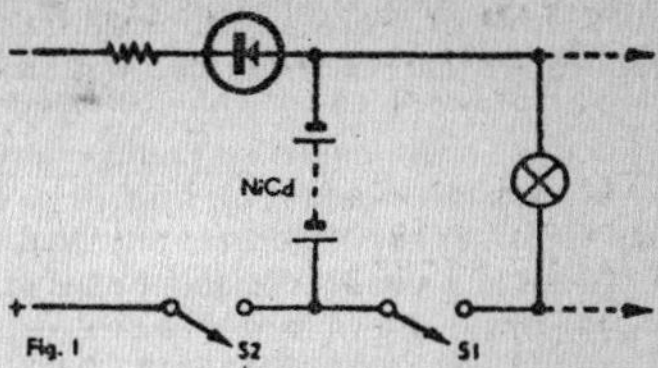

Automatic operation of S1 and S2 is a major requirement, if complete versatility it to be provided by the system.

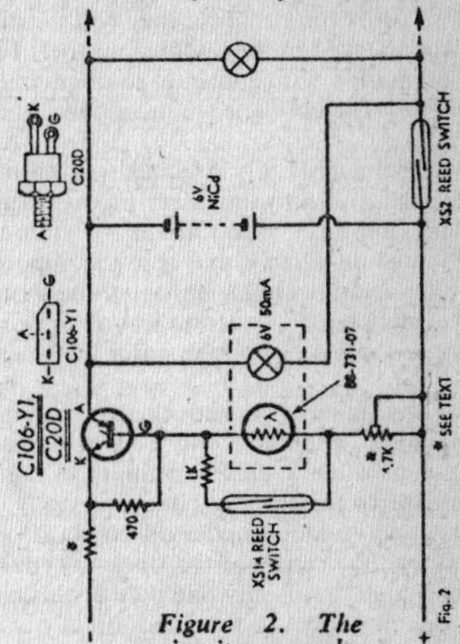

Figure 2. The circuit we evolved. The thyristor replaces the diode and also assumes the role of S2. Its gate circuit is controlled by an LDR optically coupled to a lamp across the lighting circuit.

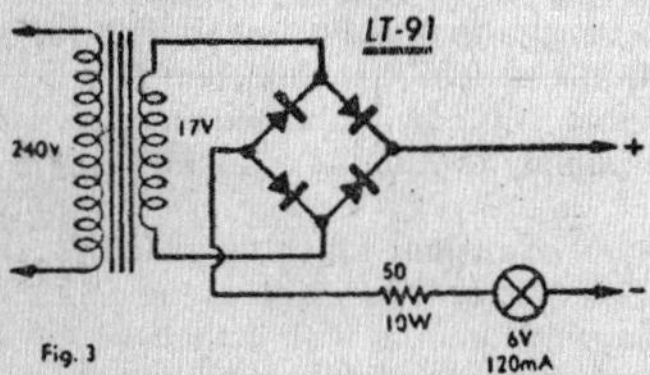

Figure 3. A typical charger circuit for use with train lighting systems. Main feature is the indicating lamp as part of the current controlling resistor.

For this reason, and also to minimise the possibility of leaving the lights on when the train is stored, some form of track-operated switch would be ideal. This would be arranged to switch the lights on whenever the train was moved out of the siding on which it was normally stored, and to switch the lights off again when the train was backed on to the siding.

No doubt many modellers could evolve such a switching system on a purely mechanical basis and if so, there is no reason why it should not be employed. On the other hand, we were intrigued by the possibility of using dry reed switches and small, permanent magnets. Using what is called a "latching" arrangement, we were able to evolve a system which works particularly well.

A latching switch is made by fitting a small magnet in close proximity to a reed switch and so spaced that it is almost, but not quite, capable of closing the reed contacts. If we now approach this combination with another magnet, so orientated that it assists the field of the latching magnet, the reeds will close. More importantly, they will remain closed after the auxiliary magnet is removed, held in the latching magnet. To open them again, we approach the combination with a magnet so orientated as to oppose the field of the latching magnet. The contacts will now open, and remain open.

A refinement of this idea makes it sensitive to the direction of travel by the reed relative to the auxiliary magnet. Thus we can arrange matters so that every time a car moves in one direction the lights will be switched off, and switched on when it moves in the opposite direction. It is a very reliable arrangement; so reliable that, if one should try to trick it by moving a car in the "switch-off" direction when the lights are already off, it will deliberately switch them on briefly in order that it can switch them off!

The reed is slung beneath the car, running parallel with its length. The latching magnet need only be a small one, just strong enough to close the contacts when in close proximity to the switch. It is positioned as close as possible to the switch, without actually closing it. The auxiliary magnet is mounted beside or under the track, at right angles to it and offset so that one end is in the centre of the track or beneath the switch. It is important that the switch "see" only one pole of this magnet. If the switching sequence is wrong in respect ot the direction of travel, simply reverse the track magnet.

The only snag with this scheme is the amount of current which the switches will handle. Although rated at 250mA DC, the larger types, such as the XS2, appear to be capable of handling about twice this amount with little risk of sticking. Assuming that the load can be kept within this limit, which is not unreasonable for a 6V system, there should be no problem.

If the load must be larger than this it may be possible to solve the problem by providing more than one switch and sharing the load between them. Unfortunately, this may involve additional wiring between cars, unless the battery car is near the centre of the train, which is not always convenient. However, even with the battery in the last car some separation can be provided by arranging that the battery car lights be controlled by its own switch and that the rest of the train be controlled by a second switch in the next car. The alternative is to provide an individual switch on each car.

This covers the broad aspects of the suggested scheme.

Following preliminary experiments, we set out to construct a train lighting system which we felt would be as nearly representative as possible of the kind of thing a typical enthusiast would require. What follows is based on the problems we encountered and how we overcame them.

The project we set ourselves was to produce a complete lighting system for a five-car train, based on a 6V battery composed of five Eveready C900 nickel-cadmium cells. On the basis that such a battery could be discharged at 900mA, this would allow just under 200mA per car. Assuming the availability of lamps in the 40mA to 50mA range, this allowed for four to five lamps per car.

The lighting system was to be controlled by means of a reed switch beneath the car, operated from magnets beside or under the track, as described in detail already. It was also felt that we should try to provide a system whereby the train could be run "lights on" or "lights off" at will, thus satisfying the requirements of simulated nighttime or daytime conditions respectively.

Finally, some means had to be provided whereby the battery could be connected to an auxiliary charger, via the pickup bogie and a short length of special track. This is necessary to offset any abnormal current drain which cannot conveniently be made good during running time.

By the time we had listed all these requirements, we realised that what had started out as a simple idea was rapidly assuming quite complex proportions. In order that it remain within the constructional capabilities of the average modeller, we would have to satisfy all these requirements without increasing the complexity of the basic circuit to any significant degree. We feel that we have achieved this particularly well.

In order to follow the development of the final system, let us start with the basic circuit (figure 1), similar to that given previously. It consists of a nickel-cadmium battery, a diode to act as a "cut-out", a resistor to control the rate of charge, and two switches, one (S1) to control the lights and one (S2) to control the charging circuit.

Assuming for the moment that these switches are manually operated, and that there is no objection to this, then such a circuit can satisfy almost any likely requirement. The train can be run "lights on", in which case both switches are closed; "lights off", in which case both switches are open; it can be stored ("light's off"), in which case only the light switch need by opened; or it can be stored ("lights off") with an auxiliary charger connected, in which case the lighting switch should be open and the charging switch closed.

However, as soon as we consider the use of track operated reed switches, as discussed earlier, we are presented with both a number of advantages and a number of problems. The main advantage is that the lights can be switched on automatically as the train leaves its siding and off again in the same manner when it is returned to it for storage. In addition, a number of interesting possibilities present themselves by using electro-magnets at various points in the system, such as at the entrance to and exit from tunnels, whereby the lights may be switched on and off in an appropriate and realistic manner.

The disadvantage is that we may not be able to avail ourselves of these possibilities, by reason of the need to provide an equivalent of S2 and, more importantly, make it do-all the right things according to the particular requirements at any one time. For example, a circuit without S2 would have to be run "lights on" at all times (to avoid overcharing the battery), but this may not always be appropriate and, what's more, prevents us from exploiting other track switching facilities, such as at the tunnel entrance and exit. While it may be possible to operate two reed switches side by side, both in a latching configuration and operated by the same track magnet, such an arrangement would, at best, be critical and difficult to adjust.

Our effort to overcome these problems is shown in figure 2. It is a little more complex than figure 1, but its operation is relatively simple, and a logical step from the basic circuit. We don't think you will find it hard to follow.

First, in place of the diode of figure 1 we have substituted a thyristor. This device can be made to function in the same manner as the diode but, in addition to the normal anode and cathode of the diode, it has a third electrode called a gate. By applying suitable voltages to the gate, relative to the cathode, we can control the current flow between anode and cathode. As far as this circuit is concerned, the thyristor is performing two functions: that of the diode and the switch S2 in figure 1.

To make it perform in this manner the gate is first connected to the cathode via a 470 ohm resistor. Under these conditions the thyristor will not conduct and "S2" is open. To make the thyristor conduct we must make the gate positive with respect to the cathode, which we do by connecting it to the positive side of the circuit via another resistor. If we make this resistor variable we can vary the thyristor operating conditions between conduction and non-conduction. When the resistor value is low the gate is positive and the thyristor conducts ("S2" closed). When the resistor value is high the gate is close to the cathode potential and the thyristor does not conduct.

Now comes the cunning part. The variable resistor we spoke of is not a conventional variable resistor; it is a Light Dependant Resistor (LDR) or a resistor which varies its value according to the amount of light falling on it. Under no light conditions its resistance can be very high—of the order of several megohms. Under bright light conditions it may be as low as a few hundred ohms.

For our purpose we couple the LDR—optically—to a lamp wired to the train lighting circuit. Under these conditions the thyristor is able to sense whether the lights are on or off. When they are on it will conduct and when they are off it will not. In practice, a very small lamp, such as a 6V, 40mA type, will supply enough light to operate the LDR. The lamp and LDR assembly is wrapped in black plastic tape to prevent false operation due to ambient light.

The result is a lighting system which can be switched on or off at will, without fear of damage to the battery, since the charging circuit will automatically follow suit.

This leaves only one requirement to be satisfied: that of auxiliary charging. And, since we have deliberately coupled the charging circuit to the lighting circuit, so that the one cannot function without the other, it appears that we must provide an auxiliary switch to by-pass the LDR when charging is necessary.

This is most easily provided by one of the small reed switches, such as the type XS14, located under the roof of the car and operated by a magnet suspended above the track and just high enough to clear the car roof. The switch is connected across the LDR but in series with a 1,000 ohm resistor. The purpose of the latter is to simulate the resistance of the LDR, when illuminated, and which normally serves to limit the thyristor gate current to a safe value. The resistor can be a fractional wattage type.

As we envisage a practical set-up, charging will be provided from the track on the siding where the train is normally stored. A convenient arrangement is to make the last car the battery car with the rear bogie providing the pickup from the track. A short length of track at the end of the siding is insulated from the remainder, and connected to the charger. The light switch magnet is so placed that it opens the reed switch shortly before the rear bogie enters the charging section, while the rooftop magnet closes its magnet shortly after the bogie enters the charging section.

In order that the operator can tell when he has placed the car correctly underneath the magnet, it is essential that the charger be fitted with some means of indicating that charging current is flowing. A practical charger circuit is shown in figure 3, and the values shown are based on the requirements of the C900 cells. It is a constant current-type charger, with a fairly large value of series resistance to control the charge rate. To provide a charging current indicator, we simply replace part of this resistance with a small lamp, in this case a 6V 120mA type.

Coming to some of the more practical problems, we have to decide how best to control the charge rate of the battery while the train is running. In small systems involving a total current drain of around 200 or 300mA a simple series resistor, as shown in figure 2, is a convenient and satisfactory method of control. However, in a system such as we are considering now, the total current to be supplied to the light-system is approximately 1A, and the heat requiring to be dissipated by the resistor can prove an embarrassment.

The problem is not so great where a metal car is used, since the resistor could be mounted against the metal and use the whole car as a heat sink. Plastic cars present the real problem, since they not only are unable to help dissipate the heat, but must be protected from it.

Where a resistor has to be used, it is best slung beneath the car where it can operate in "free air" conditions, and even benefit a little from the movement of air past it as the car moves.

However, by introducing resistance into the gate circuit of the thyristor (in addition to that of the LDR) we can control the period through which the thyristor conducts during each pulse of power, effectively varying the average charge rate.

Figure 2 shows both arrangements, but they can be used separately or together according to requirements. For small installaitons or where a metal car is used, the resistor alone should suffice. Where a plastic car is used medium installations may need both, if it transpires that the range of control provided by the thyristor is insufficient. For large systems, the thyristor alone seems the best arrangement.

However, for the latter arrangement some protection must be provided against excessive peak repetitive current. In general, the natural internal resistance of the average power supply, including our own supplies described in February and March this year, is quite sufficient to provide this protection. In the case of conventional supplies, operating with a variable resistance controller, it may be wise to include some additional resistance, say 1 to 2 ohms, in series with the variable resistor.

Also in the interest of peak repetitive current rating, it would be wise to choose a large thyristor than might be indicated purely on the average current requirement. Where series resistance is used, either wholly or partly, to control the charge rate, a thyristor such as the C106Y1 with a rating of 2A should be quite adequate. Where the amount of current dictates dispensing with the resistor, the C20D rated at 7A is a better proposition. Ideally, this should have a small heat sink, such as a metal car, or a strip of aluminium in the case of a plastic car.

We used a strip of aluminium 7/8in wide and about 7in long. About 5in of this was accommodated against the bottom of the car, underneath the interior fittings. The remaining two inches was bent at rightangles, then at rightangles again to form the support for the thyristor, which was located at the end of the car, behind the entrance door. This amount of heat sink effectively reduced the thyristor temperature, without getting more than mildly warm itself.

Where a resistor has to be used, we suggest something along the lines of the IRC PW5. When used in a metal car its square section makes it suitable for clamping to a flat metal surface, should this seem to be necessary. For plastic cars it can be slung underneath and is shorter than the width of the car, enabling it to be mounted at rightangles to the length of the car and rendering it relatively unobtrusive.

When we came to build our system, one of the first things we had to decide was the type of lamps likely to be available, since the number of other decisions depended on this. Unfortunately we do not have access, in this country, to all the miniature lamps manufactured overseas, due to the need to indent these in rather large quantities in order to obtain a reasonable price. This is rather frustrating when one studies the wide range of lamps listed in overseas catalogues, many of which would be ideal for the job.

Nevertheless, we managed to find three types which are either already in stock locally, or are expected to be by the time this article appears. A fourth type is being investigated.

The first is a "Vitality" brand, type 673, 6V, .36W, E5.5 base, about 5mm diameter and 15mm long. The .36W rating converts to a current of 60mA, but the samples we tried were closer to 45mA, which is all to the good. We understand that variations of this order are quite common with small lamps.

This lamp is excellent for general purpose carriage lighting, where its small size and modest current consumption makes it possible to light multiple compartments unobtrusively and without excessive power.

The second lamp is the Taunuslicht type T4.6, an ultra miniature version of the well-known switchboard indicator lamp, as used by the P.M.G.'s Department. It measures 4.6mm diameter, 22-mm long, and is fitted with flat contacts on the side. These should not be difficult to remove if this is desirable. It is rated at 6V, 50mA, and would be suitable for applications similar to those mentioned for the vitality lamp.
These are a premium quality lamp, by reason of their intended application, and should have a long life. They would be preferred in any situation where replacement was difficult.

Another lamp in the same brand is the T5. This is similar electrically, but is 5.4mm diameter, and 30mm long. In many cases the additional length will be of little consequence, while the diameter increase is less than 1mm.

Even more interesting is another version of the T4.6, rated at 6V 20mA. This would be ideal where it was necessary to light a large number of small compartments, while keeping the current drain within reasonable limits. This lamp is not available in Australia at the time of writing, but the importers are anticipating that local demand will justify a shipment in the near future.

By far the most intriguing lamp we discovered is the "Alba" type 10, rated at 6V, 50mA. It measures only 2mm diameter, 7mm long, and is equipped with flying leads. While suitable for general lighting, its size makes it ideal for such special applications as the small marker lights on the end of cars, fitting into existing loco headlights etc. If it is felt that the amount of light is too great for marker lights, two lamps may be connected in series across a 6V system, and still give useful light output. In these circumstances two lamps may be operated for a total current drain of about 35mA.

When setting up a lighting circuit for final adjustment, we suggest that the car be placed on a short length of dummy track, connected in parallel with the main track. The latter should form a complete circuit, so that the locomotive can be set running on it, its speed set to "average", while adjustments are made to the charge rate, etc., in the stationary car.

A milliamp meter should be connected directly in one of the battery leads–whichever one is most convenient will do–in which position it will indicate both charge and discharge conditions as a defection either side of zero. By selecting the right polarity, it can be made to read up the scale for the charge condition, which is what we wish to observe. By the time the system is adjusted to provide the correct charging current for the battery, 90mA in this case, the main power supply is, in fact, already supplying all the power needed to operate the lights, plus the charging current for the battery.

The reed switch for the lighting control was a type XS2, as discussed earlier. Although the current involved was nearly 1A, and about four times the figure quoted for maximum life, we experienced no problems and no signs of sticking, such as occurred with out original single-cell system. Naturally, such an order of overload cannot have the blessing of the manufacturer, and must shorten the life of the reed. However, in this simple application, we feel that such a liberty is permissible, and that the shortened life might still be too long to worry anyone.

Coming to the more practical side, we face the job of packing the nickel-cadmium cells and the associated components into a single car, while retaining some semblance of interior fittings. We chose a combined brake van and coach, with the idea of locating as much equipment as possible in the brake van section, which has fewer windows, Such windows as there are were "frosted" by the simple procedure of fastening a strip of plain white notepaper behind them. This section may still be lit–using two "Alba" lamps, one on each side of the battery–which will help to disguise the real function.

we located four of the five cells in the brake van section, together with a small panel of Zepher board which accommodated the LDR and its associated lamp, the trim pot to control the charging rate, and also served as a terminal panel. The fifth cell was located in the passenger section, against the roof, and is visible only when the car is near eye level. At the normal angle at which models are viewed, it is completely out of sight.

In wiring the car it is not recommended that leads be cut to minimum length, otherwise maintenance and adjustments will be unduly difficult. Make each lead into a neat but reasonable size loop which will allow sections to be moved if necessary.

The next problem was to provide a simple and aesthetically acceptable method of connecting all the cars together electrically. We approached this on the basis that the train would normally remain complete and not require to be broken up while in use. On the other hand, the connections should be capable of being broken is it was necessary to remove a car for repairs.

Our final solution was to fit each car with a pair of contacts at each end, into which plugged a small half loop of light hookup wire which could pass as ar air hose. The contacts are from a seven pin or nine pin valve socket and are cemented underneath the car with one of the epoxy compounds. Access to the contacts is by means of a small (3/64) hole drilled in the end of the car.

Pins for the sockets were fabricated from wire paper clips, which are the right gauge and suitably stiff. Each pin is "L" shape, the long side being about 3/16in and the short side about 1/16in. The loop of hook-up wire is about 1¼in long and is fitted with a pin at each end. The wire is soldered to the short side of the pin.

When cementing the contacts under the car take care that the cement does not extend so far back or is so bulky that it fouls the tops of the front pair of bogie wheels. We found this out the hard way!

Lighting a multiple-compartment coach presents something of a problem since, at first glance, one is required to provide a lamp for each compartment, which may total seven or eight. Unless lamps in the 20mA range are available, this is more power than can be spared for one car. We overcame this by dividing one 40mA lamp between two compartments. The lamp is mounted horizontally against the roof in a small cut-out in the partition dividing the compartments, with the lamp filament on the line of the partition. This will share the light equally between the two compartments, and works particularly well.

Finally, a few words about the power pick-up bogie. The individual reader's approach to this will vary, according to the type of car involved and the workshop facilities available, but our approach may provide some ideas.

Our coach was a Triang, fitted with plastic wheels mounted on pointed (needle bearing) axles, the bearings being moulded in the plastic bogie assembly. We found it quite simple to remove these wheels, so we passed one over to our engineering department and had them turn up two brass copies. However, if a ready-made metal wheel is available which is a push fit on this axle, it may serve as well.

After confirming the position of the brass wheel on the axle to maintain the gauge, we removed the plastic wheel while we soldered the brass wheel to the axle. Contact with the wheel is through the axle and the photograph shows how the two contacts are fitted. These can be made from a light-gauge spring brass. They are secured to the plastic bogie with 10BA brass screws, which also retain a soler lug from which leads run into the car.

Solder the leads to this lug before it is mounted under the screw head, otherwise the heat will damage the plastic.

The car we used was the assembly pack version, rather than the ready-made one, and differs in that the bogie is held with a stud and nut rather than a rivet. This is important where the bogie has to be removed. Where the stud and nut is used, it saves a lot of time if the nut is secured in its normal place with a spot of glue or epoxy. This makes it possible to detach the bogie without the need to pull the whole car to pieces.

In our discussion on train lighting systems so far we have mentioned both permanent magnets and electromagnets for use in conjunction with reed switches, but we have not had the opportunity to discuss them in detail.

The permanent magnets used in our experiments were supplied by the Rola Company, and are of the type known as "stick-cast" magnets. These are cast in large blocks, with each magnet separated from its neighbout by a notch which enables it to be easily detached. They are available in a variety of sizes, ranging from 3/8in long to nearly 2in long, are supplied ready magnetised, and are relatively inexpensive.

The largest size, type FM448, is 1-15/16 x 13/32 x 3/16in and is particularly useful for our purpose. It is strong enough to operate a reed switch under the car when placed beside the track, or under the roof of a car when suspended about ½in above it. We have also found it suitable for another application with which we have been experimenting; reed switches in the track operated by magnets on the train to operate signalling and interlocking circuits. We will have more to say about this in a later article.

The next smallest sizes, types FM605 (1-1/16in long) and FM464 (7/8in long) are useful where they can work in closer proximity to the reed switch. One such application is where it is convenient to mount the magnet directly under the track, vertically, with one end no further away from the reed than the underside of the sleepers. Better still, sleepers may be cut away to bring the top of the magnet flush with the top of the sleepers.

At the other end of the scale we require a magnet of only limited strength for the latching function. Initially, we tried a type FM176, measuring ¾ x 3/16 x 1/8inch, but this proved too strong. In some cases this can be overcome by spacing the magnet away from the reed by a small amount, but this is not always convenient. We tackled the problem by breaking the magnet in half—this requires nothing more than a vyce and a pair of pliers—and this turned out to be just right.

Later we had the opportunity to try the type FM626, the smallest of the series (3/8 x 3/16 x 1/8 inch). This also appears to be quite satisfactory and is probably more convenient than having to break up the larger type.

While permanent magnets are ideal for those applications where the same function is always to be performed at the same point on the track—as, for example, switching the lights off when the train is shunted on to its storage siding —they are less suitable where requirements vary. It may be possible to use them if the modeller has the skill and facilities to provide mechanical linkages from the control point to the magnet, whereby the latter may be moved out of operative range or completely rotated for a reverse function, but most people would prefer to avoid such complication.

Electromagnets, on the other hand, offer the desired versatility with no more complication than running wires to the various points at which switching is required. As we see it, there are two broad applications for the electromagent; "once-only" switching at the decision of the operator, and semi-permanent switching occurring automatically for as long as may be required.

In the first category we have the situation where it is required simply to switch the lights on or off and leave them like this until a new condition indicates a further change. Thus, if the train is taken out of the siding and the lights automatically switched on, it may then be necessary to switch them off if lights are inappropriate to the particular display. Thus the first place to locate an electromagnet is alongside the main track just beyond the points.

The seond category would be provided by a tunnel in the system, In this case, when running the train "lights off", we require to switch the lights on as the train approaches the tunnel and off again as it emerges from the other end. And, since the train – or trains – may pass through the tunnel several times in the course of a display, we would prefer to have these switching functions operate automatically, i.e., by simply putting a switch marked "tunnel lights" to the "on" position and leaving it there the same sequence would occur every time the train passed through the tunnel.

If, subsequently, we wished to run the train "lights on", the same switch must be set to the "off" position to prevent the lights being turned off as the train leaves the tunnel. This is the main reason why permanent magnets cannot be used in this application.

When we came to consider the practical aspect of providing electro-magnets, we were on the lookout for some ready made devices which would avoid the need to wind a magnet. During this search we came across some old vibrator cartridges discarded by our mobile radio maintenance mechanic and immediately considered the possibility of using the reed energising coil. At least it would give us some idea of what was likely to be required.

We stripped down a cartridge, noting in the process that it contained some useful nuts and screws, insulating washers, spacers etc., plus lengths of springy metal which might be useful in making pickup bogies, as described earlier. The coil is mounted in a "U" shaped metal bracket, which we removed by cutting through the bottom section of the "U" each side of the rivet head holding the coil. We don't recommend grinding the head off the rivet, since this may release tension on the ends of the coil and allow it to collapse.

Power for the coil was taken from the charger (figure 3,) which, while the train is running is serving no other useful purpose. We added resistance in series with the coil to keep the current drain within reasonable limits, mounted the coil alongside the track, and set out to find the minimum value of current which would operate the light switch reliably.

These coils appear to vary quite markedly, some working as low as 300mA, others requiring as much as 500mA. It would seem that one would have to allow for the latter figure to be on the safe side. Such a coil, connected to a 3-position switch wired to give reversed polarity either side of the central (off) position, can be used for once only switching quite satisfactorily.

However, at this current drain the coil gets fairly hot after a few minutes running, and so is unsuitable for the semi-permanent functions we envisaged. Some improvement is possible if the coil can be mounted under the track, since this reduces the distance between it and the reed and allows it to operate at lower currents. This method of mounting has the advantage that the coil is hidden and does not have to be disguised, but the disadvantage that it may mean taking up the track, not always a convenient proposition.

For once-only switching there are several possible arrangements, but a common aspect would seem to be the provision of as many coils as possible along the track so that a train is never very far from one of them, thus allowing the light switch to be operated with no more than a few seconds delay.

Assuming enough power was available it would be most convenient, from an operational point of view, to simply connect all the coils in parallel via one polarity reversing switch. Then, to perform either switching function, one would simply set the polarity switch to the appropriate position until the train passed the next coil, then restore it to "off". The main objection is the power required, which would amount to several amps for a typical layout.

The alternative is to switch power to each coil as required, but this presents the objection that two switches, would be needed, selector and polarity, and both of these would have to be manipulated while the train was moving. Selecting the right coil to suit the train's position on the track might not be easy, unless it could be coupled to a block system which indicated the section of track in use at any time.

It was while we were pondering on these problems that we came up with another idea. If the electromagnet could not do the job unaided, at a reasonable current drain, why not give it a helping hand by the addition of a permanent magnet? A quick trial run produced a quite surprising result. A coil which needed over 450mA for unaided operation was able to produce completely reliable results at 110mA by this system. At this current, a coil could be left energised indefinitely and never get more than warm.

There is one catch: This arrangement can work with only one coil polarity, meaning that, for any one direction of travel, it can perform either an "on" function or an "off" function, according to the polarity of the permanent magnet, but not both. Nevertheless, it is ideally suited to our tunnel switching requirements, which simply requires a predetermined switching pattern that can be turned on or off as required.

It is less suitable for once-only switching, but one arrangement does suggest itself. This is to make each coil perform an opposite function, increasing the number of coils if this is feasible. The arrangement is quite simple, all coils being connected with the same polarity, and supplied via a polarity reversing switch. The required function of each coil is determined by the polarity of the permanent magnet associated with it.

Having thought of the idea, we mocked up a relatively simple version of it on our own experimental layout. This is an oval pattern, approximately 6ft x 4ft, with one siding to store the train, and we set up four switching points approximately equally spaced around the track. Two were designed to switch off and two to switch on, those with the same functions being diametrically opposed. One of the "off" positions was deliberately located just beyond the junction of the siding with the main line, providing an opportunity to switch off the lights immediately the train leaves the siding, normally with the lights on.

The more we use this arrangement, the better we like it. The mere fact that one can operate the train lights while the train is running is not only a novel feature, but an extremely convenient one. It is simple to use and one hundred per cent reliable. For the reversing switch we used a 3-pole, 3-position, rotary switch, but we suggest that a P.M.G. key-type switch would be a more convenient type.

Ideally it should be spring loaded in each direction, so that it automatically returns to the neutral position when it is released. It should be mounted so that its movement is parallel to the track and so wired that it performs the "on" function when moved in the same direction as trains normally run on that track, i.e., "east" or "west", "up" or "down", according to the terminology you employ. When pointing against the direction of traffic, it performs the "off" function. This system makes it easy to remember the switch positions.

At the low current required for this system it becomes feasible to connect several coils in series, thus minimising—or even eliminating—the wasteful series resistance and keeping the current drain to a minimum. The coils have a resistance of about 15 ohms, but their impedance is higher than this when supplied with unfiltered current from a rectifier. We were able to operate at least six coils in series without any filtering, while the addition of a couple of hundred microfarads across the rectifier increased the current flow significantly. We imagine nine or 10 coils could be operated in this manner.

In any case filtering may be desirable, since at least some vibrator coils are designed with an auxiliary shorted ("damping") winding associated with the main winding. These would generate heat when associated with anything other than a steady field.

The permanent magnet used to aid the electromagnet must be of such a strength, and so placed, that it is just incapable of operating the reed. When the extra field of the electromagnet is added the reed will function. We located the permanent magnet on the opposite side of the track from the coil, the correct polarity for the particular function being determined experimentally. One of the medium size magnets, such as the type FM464, is suitable for this job.

Unfortunately, for our purpose, vibrator cartridges are not as plentiful as they once were, most of their jobs having been taken over by solid state devices. Nevertheless, we suggest you contact your local serviceman and advise him that you are in the market for any discards. Any he acquires he will probably be only too happy to get rid of!

The alternative would seem to be to wind one's own. To get some idea of what was involved we first stripped down a vibrator driving coil and noted its dimensions, gauge of wire, etc. The former is of metal throughout, though covered with insulating material over the winding areas. The core is a flat strip of mild steel ¾in x ¼in x 1/16in, the cheeks measure ¾in x ½in and one is of mild steel and one of brass. The brass end is the business end of the magnet. The distance between the cheeks (inside) is 9/16in.

The driving coil is wound with 34SWG enamelled copper wire. We had some difficulty unwinding it, due to its being interwound with the damping winding, and we were unable to count the turns. However, if a former of approximately this size is wound with 34 gauge wire to a resistance of between 12 and 15 ohms, it should be close enough.

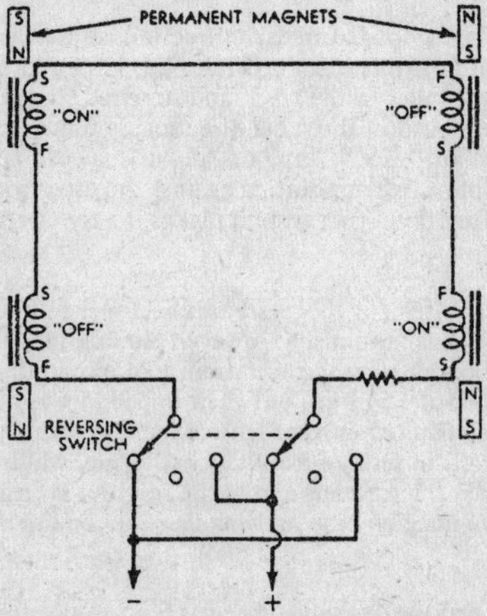

A typical arrangement of electromagnets connected in series and fed via a reversing switch. Power can be taken from any convenient auxiliary supply and the resistor is selected to give minimum current consistent with reliability.

MODEL TRAINS – SIGNALLING AND INERTIA

Following the completion of our train lighting system, we turned our attention to the problem of extending the benefits of simulated inertia intc automatic signalling and interlocking systems. The result is two simple circuits which can be added to our simulated inertia control unit to convert them to this application.

Our Model Train Control with Simulated Inertia was developed out of a desire for more realistic acceleration and deceleration in model train systems. It was aimed particularly at the outrageously unrealistic situation where a train enters a platform or approaches a danger signal at anything up to a scale speed of 60 miles an hour, yet comes to a dead stop in half a turn of the driving wheels. Pity the poor passengers!

Our control unit does a particularly good job in this regard, even though it is a little disconcerting at first to realise that a train cannot be stopped in less than several train lengths. However, once one gets used to this characteristic, and learns to "drive" the train as a real driver would, results are both realistic and gratifying.

But then we faced another problem. Assuming that a layout has any kind of automatic signalling or interlocking system, whereby the presence of a train in a section automatically kills the section behind it, how would this fit in with our artificial inertia? Since the inertia is really in the control unit, and not the train, a train encountering a dead section of track can still come to an abrupt halt from 60 miles an hour.

What was needed was some method whereby any automatic signal would cause the train to slow down and stop in accordance with its inertia, then start and pick up speed in the same manner in response to an all-clear signal, all without the intervention of the operator or without disturbing the setting of the main throttle control.

Studying the original artificial inertia circuit, we realised that there was a natural point of entry into the system between the OA91 diode and the braking resistors. We broke the circuit at this point and wired the two halves to a pair of terminals on the rear of the control box. Simply opening the circuit at this point would be equivalent to setting the throttle to zero, and this alone would bring the train to a gradual halt, though probably over too great a distance. But if, at the same time, a suitable value of resistance ("auxiliary brake") could be switched across the existing brake circuit it could provide fixed amount of braking. By selecting a suitable value, the train could be brought to a stop in a reasonable distance.

Both these actions, throttle and brake, could be provided by a change-over set of contacts, as shown in the modified circuit. Typically, this contact set could be provided by a relay, and any impulse which operated the relay would cause the train to come to a halt in a gradual and realistic manner. At the same time, the main throttle and brake controls on the control panel would remain as they were set, ready to take over when the relay was released. The train would then simply accelerate in accordance with its inertia until it reached the speed originally set by the throttle.

Taking the idea one step further, it obviously would be more realistic if the train were to be halted at a predetermined point, such as a signal at the entry to the next section, rather than at some arbitrary point dictated by the moment at which the next section was occupied.

This meant providing some form of sensing device in the track to energise the relay when the train passes over it, and which could be located some distance down the track from the signal or stopping point. The most logical choice was a dry reed switch and the most logical type the miniature XS/14. For our experiments this was simply rested on top of the sleepers, running parallel between the rails, and this may suffice if it can be disguised in some way. Alternatively, the sleepers could be cut away to allow it to be lowered and covered with ballast.

The reed is operated by a magnet mounted under some item of rolling stock in the train, preferably the loco. We found it most convenient to mount it under the tender, simply glueing it to the underside. However, if the reed is recessed to near sleeper level, it may be necessary to lower the magnet by spacing it away from the underside with a suitable thickness of cardboard or plywood. This would bring it to a level just above the axles. We suggest one of the larger magnets for this job, such as the Rola FM448. As will be seen, it is important that the reed be held closed for a brief period, even if the train is moving at maximum speed.

This reed is connected in series with the relay coil, relay power supply, and another set of contacts which can be anywhere in the interlocking or signalling system. In our case we simply made these contacts part of a signal so that when the signal was up (danger) the contacts were closed and when it was down (clear) they were open.

Finally, we had to make some provision to lock the relay in after the train had passed over the reed switch. This was done by using an extra set of contacts on the relay, normally open, and wiring them in parallel with the reed. Thus, when the relay pulls in it locks itself in with a pair of its own contacts.

Now let us consider the sequence of events. The signal is set to danger as the train approaches, which leaves the reed switch as the only open circuit in the relay coil circuit. As the train passes over the reed it closed momentarily, which must be long enough for the relay to pull in. When the relay pulls in it locks in and operates the changeover contacts which close the throttle and apply the local braking circuit. The train then comes to a gradual halt in a distance determined by the local braking resistor value.

When the signal is lowered the circuit to the relay coil is opened, the relay drops out, and the original control panel settings are restored. The train

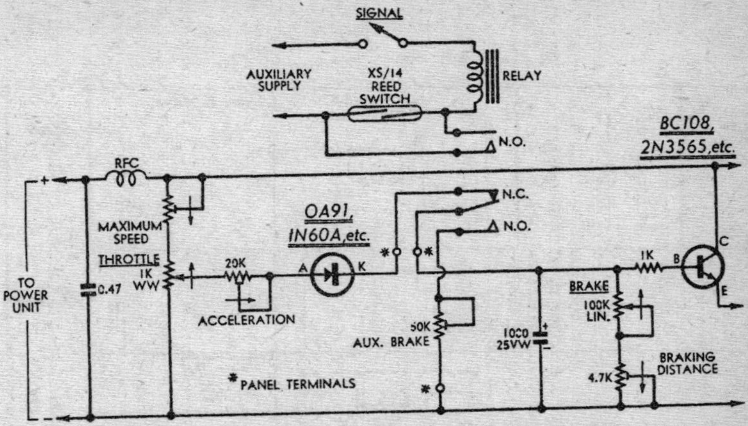

A simple system of connecting a signal or interlock circuit into our Inertia Control Unit. Only a portion of the latter circuit is shown

The simple circuit on the the previous page can be converted to this version with only a few additional components. It is preferred when it is required to stop a train at a precise position, such as at a signal or in a station

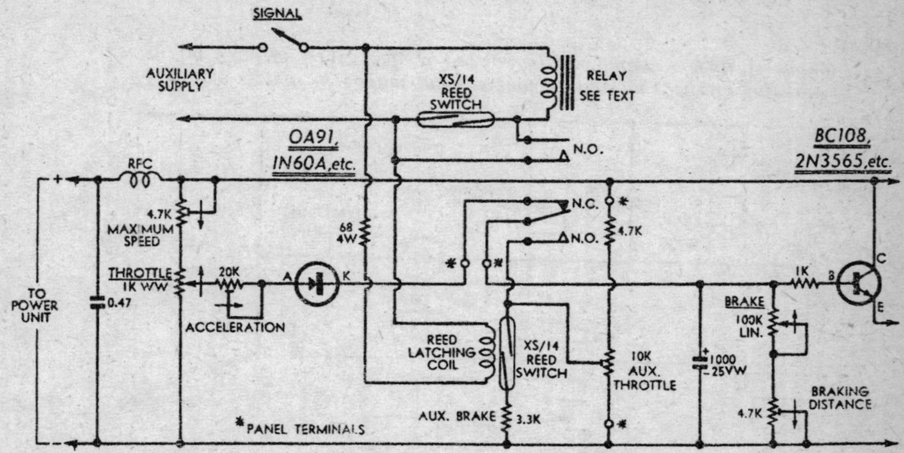

may hesitate for a second or two – a good time to sound a whistle or hooter if one is available – then will move off slowly and pick up speed in the normal way. If it should transpire that the signal is lowered after the train has passed over the reed, but while it is still slowing down, it will simply pick up speed again in accordance with its inertia, in exactly the same way as would a real train.

We located all the extra components – relay, auxiliary brake, and so on, outside the main control box, connection being via the two extra terminals already mentioned, and the negative line terminal, whichwas already fitted. The extra circuitry and components can be housed in any manner which is convenient.

The relay is a standard unit, having the contact arrangements already described, and fitted with a 200ohm coil. Power for the relay comes from the same supply which we use to charge the train batteries and operate the trackside light switching magnets. It can easily cope with these short term current demands while not charging the batteries. The relay coil resistance is not critical, but should be kept fairly low to ensure that the relay snaps in smartly when the reed is energised, since it must close its own holding contacts before the train passes and releases the reed.

The overall effect of this set-up is particularly realistic, especially if one can manage to completely hide – rather than disguise – the reed switch. In such circumstances the effect is not only realistic but quite puzzling to those enthusiasts who see it for the first time. And, while some refinements are possible, even as it stands it is well worth trying. In any case, we suggest that it be built as a first step anyway, since none of the work or components will be wasted when it is decided to progress to the improved version.

A limitation of the first version is that the stopping distance of the train is dependent on its speed. If the circuit is adjusted so that, at a typical maximum speed, the train stops at the signal, it will stop short of this point at all lesser speeds, or will overshoot should the "typical maximum" be exceeded.

In most cases the error is not large but, should the train happen to be travelling very slowly, it may come to rest in half the distance or less than that required at a typical maximum. We dedided to see if we could overcome this.

It wasn't hard to come up with some ideas, but most of them tended to be too complex for what we were trying to achieve. Finding a simpler way to do it took most of our effort, but we are very happy with the end result, both in terms of performance and circuit simplicity.

Basis of our thinking was to provide two track sensors, one located as before, and one located near the desired stopping point, in this case the signal. The first sensor would operate the relay as before, but the latter would be so arranged that, instead of closing the throttle completely, it would retard it to a very low speed – about the lowest speed at which the train would run reliably. The second sensor would introduce the auxiliary braking resistor, so chosen that it would stop the train at the required point, assuming that it had already dropped to the selected minimum speed.

Provided the train could be decelerated to this new speed from the maximum speed likely to be used, then all lesser speeds could be handled. The train would simply drop from its running speed to the minimum speed, then continue on towards the signal at this rate. It would therefore always be travelling at the same speed when it encountered the braking circuit and would stop in exactly the same distance.

Achieving a new throttle setting, using the existing relay, didn't seem to present many problems, but we were concerned about the additional braking circuit. It was easy enough to fit a second XS/14 reed switch in the track, and wire it as required, but we had to provide some form of latching to ensure that the brake would remain on even if the magnet in the loco passed right over it and beyond. Furthermore, the latching circuit had to be controllable, so that it could be "on" or "off" as dictated by the signal position.

Our first reaction was to use another relay, with a latching function similar to the first, but we discarded the idea almost immediately; it seemed unnecessarily complex and expensive to provide a relay simply for this function.

Using a permanent magnet as a latching device appeared awkward, since we had to have some means to unlatch it when the train moved off again. We considered that this might be done with a magnet in the last car of the train, so orientated as to oppose the latching magnet. Such an arrangement might work, but has the objection that the distance between the two magnets in the train must always exceed the braking distance provided by this reed circuit. This would be no problem in a complete train, but might be difficult to provide in a loco alone, particularly the smaller tank variety.

Finally we hit on the idea of using an electromagnet for the latch, and in a particularly simple form. Why not simply wind a coil directly on the glass tube of the reed switch? If we were lucky we might not need to provide more than two or three winding layers. Reference to the data sheets showed that the XS/14 required 20 ampere turns for the "hold" condition, so we set to work to work out how this might be achieved.

The finest wire we had on hand was 34B&S enamelled and a winding table showed this as giving 144 turns to the inch. With a winding length on the glass of a little over 5/8in we calculated that we could get close to 100 turns in a single layer. Even if we stopped at that, a current of 200mA would provide the required 20 ampere turns, and this was not an unreasonable amount of current.

On this basis we made up a trial winding. We didn't bother to count the turns, but simply packed as many as we could into a single layer. The start of the winding was anchored temporarily around one of the reed terminals and a piece of sticky tape cut in readiness to secure the finish. When completed we checked it at 200-mA and were gratified to find that it latched quite reliably; in fact, it worked down to 180mA.

Since we could provide this order of current quite easily from the auxiliary power supply, through a suitable limiting resistor, we didn't bother to try adding more layers. One objection to the latter is that it would make the reed unnecessarily bulky and more difficult to accommodate on the track. For the same reason heavier gauge wire is to be avoided, but lighter gauges could be used, if available and if you feel capable of handling them.

The modifications to the throttle circuit are relatively simple. We have provided an auxiliary throttle potentiometer to which the inertia capacitor (1000uF) is switched by the relay. As with the main throttle, the position of the moving arm determines the speed of the train, but in this case we set it for lowest practical minimum speed. The total value of the pot determines the rate of deceleration from the running speed to the minimum speed; the lower the value pot the more rapid the deceleration. In our case a 20K pot was too large and 10K too small, so we added the 4.7K reistor shown.

The second reed switches in the auxilliary braking resistor, 3.3K in our case, which reduces the throttle setting still further and simultaneously discharges the 1000uF inertia capacitor. The fact the auxilliary throttle is on the opposite side of the diode to the main throttle allows the one resistor to perform the dual function.

As with the simpler arrangement, the auxiliary components were connected to the main control unit via the terminals already provided, plus one extra one connected to the positive line after this has passed through the RF choke (RFC).

The exact values of the additional components depend on so many variables that we can give only approximate values based on our own experiments. Quite apart from the differences between different locos, power supplies etc., there is the amount of track length available to bring the train to a halt. In large layouts there may be plenty and maximum realism may be achieved, but smaller ones may have to accept some compromise. On our test layout the distance between the first reed switch and the signal is about 7½ft; probably a minimum.

This system works particularly well, and will bring a train to a stop within a fraction of an inch of the same spot, regardless of whether it is flat out or merely crawling. In the latter circumstance it is likely that the running speed is so close to the speed set by the auxiliary throttle that there will be little or no noticeable change when the train goes over the first reed; it simply keeps moving until it comes to the second reed, then stops gently. The only "silly" thing that can happen is that if the running speed is too slow, the train may actually pick up slightly after the first reed. This is an unlikely situation, and even less likely to be noticed.

HIGH POWER CONTROLLER FOR MODEL TRAINS

Here is a model train controller which should meet the needs of the most demanding enthusiast. It has adequate power for almost any type loco – singly or double heading – with enough to spare for train lighting, it provides realistic artificial inertia, and it can be coupled to signal or interlocking systems so as to retain the inertia for automatic stops and starts.

This new control unit is a logical development from our "Train Control with Simulated Inertia", prompted by our own practical experience with the original version and comments from various readers who had built it. The gist of these readers' remarks could be summed up as; "It is an excellent idea, and it works fine, except –".

"The "except" involved a number of situations, but usually boiled down to the same thing: not quite enough power. While the unit worked fine with smaller locos, or types which were reasonably efficient, it was less successful with larger models, or types which were not particularly efficient.

Other situations where more power was needed was when "double heading" (i.e. two locos in tandem) or where train lights had to be supplied in addition to the loco motor. In regard to double heading, it should be pointed out that any control system which strives to provide constant torque, by sensing the speed of the motor via the back EMF, is functioning at a disadvantage when trying to supply two motors at the same time. The presence of two back EMF values tends to "confuse" the control, with the result that the two locos may not perform as well as a single loco. This problem is much less serious when the two locos are of identical type.

Having decided that a larger unit was desirable, the next point to decide was, how much larger? The next most logical thyristor available was rated at a little over 7A, but other factors had to be considered. One was the series overload lamp, which would have to be quite large to suit such a rating, while another was the likely limitation of the power supply proper. All things considered, we decided to settle for a 5A rating, and have selected the lamp, fuse, and power transformer on this basis.

We have been using this control over the last few months while developing our lighting system, signalling circuits, etc., and this has provided an excellent proving ground. In between these other projects, we have taken time off to try various improvements suggested by practical experience. A number of these proved markedly beneficial and have been incorporated, resulting in a particularly effective unit.

Before giving details of this controller, a brief discussion of the principles involved in this and other controllers may be of interest to the enthusiast whose knowledge of electronics is only modest.

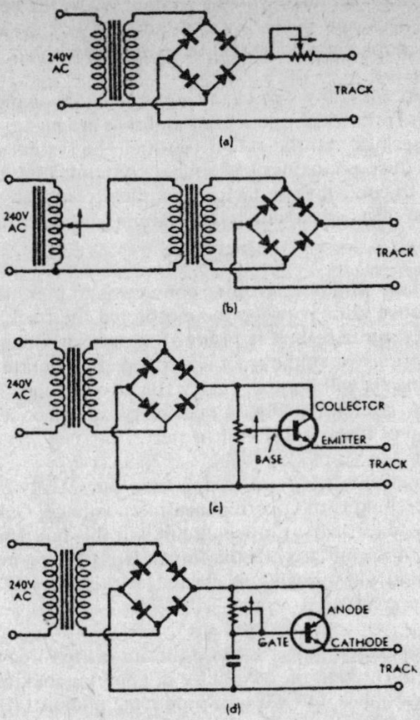

Figure 1. The four basic forms of model train controller discussed in the text. (a) Resistor controller. (b) Variable transformer or "Variac" controller. (c) Transistor controller. (d) Simple thyristor controller.

There are two basic problems to be considered when contemplating the design of power supply/control systems for model trains. First, there is the essentially practical one involving contact resistance between the loco wheels and the rails, and the loco wheels and the pickup springs; a problem which, like the poor, seems destined to be always with us.

Contact resistance is really a mechanical problem, involving such things as the materials from which the wheels and rails are made, the adjustment of the pickup springs, and the extent to which the system as a whole can be kept free from dust, oil, oxides and similar contaminants. Unfortunately, it is not possible to control these factors completely, and any aspect of the power supply system which will help to minimise them will always be welcome.

The effect of these wiping and rolling contacts is to present an ever-changing resistance value between the motor and the track, and which can be, momentarily, quite high. The result of momentary high resistance is normally not serious when the train is running at moderate to fast speeds, since the true inertia will normally carry the loco beyond this condition. However, when travelling slowly, or stationary, such a condition may cause the train to stop or make it difficult to start.

When such a condition is encountered, a train can usually be started by the "brute force" technique of applying maximum voltage. Only if the condition approaches a virtual open circuit will this fail. Unfortunately, while effective, it is quite unrealistic. Immediately the train moves the extreme condition will probably be cleared, with the result that the train takes off like a rocket.

The technique of pulsed power is based on this brute force technique, being an attempt to retain the advantage of applying maximum voltage to overcome the resistance, but without producing uncontrollable acceleration. In the ideal form, it would consist of short pulses of voltage, always at the maximum value, but varying in width according to the amount of power required from the motor at the particular moment. Thus we achieve the advantage of always operating the motor at maximum instantaneous voltage and current regardless of the average power or speed required.

In its simplest form, pulsed power may be nothing more than the unfiltered output from a rectifier, either half wave or full wave. This is better than pure DC because, for a given performance (speed), the peak voltage will always be higher than the equivalent DC value.

Since this simple form can be obtained by merely omitting a filter (which would be quite expensive anyway) in the power supply system, it is normally provided as a matter of course in all but the most elementary systems. Here, dry batteries may be used for simplicity and low first cost, but such systems are of little interest to serious enthusiasts.

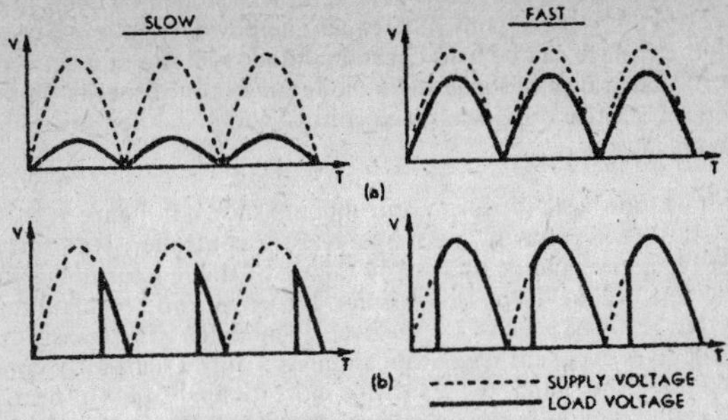

Figure 2. Voltage waveforms for (a) transistor controller (b) thyristor controller.

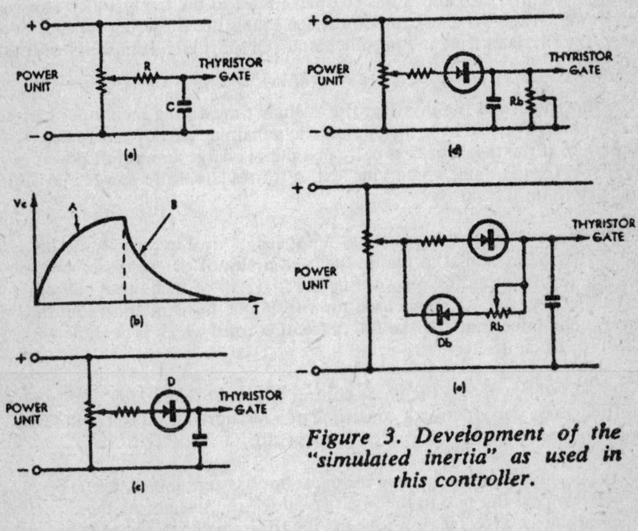

Figure 3. Development of the "simulated inertia" as used in this controller.

The second problem is to provide good regulation of the voltage supplied to the track, regardless of the current which the system may demand, or the existence of any speed control between the power supply and the track. Assuming the voltage can be held constant, a loco will run at a near constant speed over a wide range of loads and a moderate range of speeds. In the past it has been difficult to provide a speed control which did not adversely affect the regulation.

The circuit of four basic types of controller are shown in figure 1, their characteristics are as follows. The simple resistor controller, 1(a), suffers from poor regulation and possible early failure of the resistor unless it is well made. The "variac" type transformer, 1(b) gives good regulation and reliability but is too expensive for the average modeller. The transistor controller, 1(c), gives excellent regulation and needs only a light-duty control resistor. On the other hand, the transistor must dissipate a lot of heat, via a heat sink, and like the other two systems, does not fully exploit the pulsed power technique.

The thyristor control system comes much closer to the ideal. The thyristor does not act as a variable resistor, but rather as a switch in series with the track. While the gate-cathode voltage is below the device's "triggering" voltage, the thyristor appears as an open circuit to the track, which receives no power. When the gate-cathode voltage equals the triggering voltage, however, the thyristor fires and becomes almost a short circuit in series with the track.

The throttle makes the thyristor fire at some time during each input half-cycle pulse. The thyristor then passes the remaining part of each pulse to the train. If the thyristor fires early in a pulse, a large amount of power is available for the train, which runs fast. If it fires late, little power is avilable and the train runs slowly.

If a simple potentiometer is used as the speed control in the gate circuit of the thyristor, similar to that in the base circuit of the transistor controller, the thyristor can fire only during the first half of each input half-cycle pulse (90 degrees). Although used for our simple thyristor controller this arrangement does not give the full range of control which is possible with this type of controller.

A better arrangement is the phase control circuit of figure 1(d). Here a phase shift or phase delay network consisting of a resistor and capacitor in series, is connected in the gate circuit of the thyristor.

At the beginning of a half cycle the capacitor has zero impedance and effectively pegs the gate to the negative side of the system. As the capacitor charges, the gate voltage rises towards the triggering level, the time needed to reach this being determined by the values of resistance and capacitance. By selecting suitable values the triggering level may be delayed until almost the end of the cycle, or any point prior to this, enabling us to use any portion of almost the complete half cycle.

Because of its switching action, very little power is dissipated in the thyristor or in the throttle control. No bulky heat sink is required, and the control potentiometer may be a low-wattage type.

Equally important is the excellent regulation it provides, and the reason for this is worth studying in some detail. (The same theory applies to the transistor controller.)

The armature voltage of the train's motor is closely proportional to speed and this voltage appears in the cathode circuit of the thyristor. The thyristor thus has both a forward bias due to the control potentiometer and a reverse bias from the motor. Because of this there is a negative feedback action which gives very good load-speed regulation.

As the load on the motor is reduced, the speed tends to increase. The motor voltage rises, reducing the gate-cathode voltage of the thyristor. This tends to reduce the width of the output pulse to maintain an almost constant speed. The same principle applies with an increasing load and a reducing speed.

By comparison with, say, the variable transformer system which, in the ideal case (not achieved in practice), could do no more than maintain an absolutely constant voltage across the track, the transistor or thyristor system will actually INCREASE the voltage when a load is encountered and DECREASE it when a load is dropped.

When compared with the simple resistor controller, its action is the exact opposite.

The development of a simulated inertia control circuit is illustrated in figure 3. Consider first the circuit of figure 3(a). This shows the basic arrangement of resistance, R, and capacitance, C, which form the "inertia" circuit for the controller. The amount of "inertia" depends on the product of the values of R and C included in the circuit – i.e., on the R-C "time-constant."

When the speed control, or throttle, is advanced, the voltage across the capacitor does not immediately equal the voltage on the wiper of the potentiometer. Instead it rises in an exponential curve as in (A) of figure 3(b). The thyristor cannot conduct until the voltage at the gate equals the firing potential.

The time for this to happen depends on the values of the resistor and capacitor. This delays the starting of a train, and also gives a relatively slowly increasing speed after it does start. The effect resembles the inertia of a real train.

When the throttle is returned to the earth position the capacitor discharges through the "inertia" resistor. Becuase the discharge curve is the inverse of the charging curve, the thyristor ceases to fire when the voltage is still dropping at a relatively rapid rate, as in curve B of figure 3(b). Thus the time taken for the train to slow down and stop is relatively short compared with the time taken to speed up.

This is not realistic enough, and some provision must be made to slow down the discharge of the capacitor. This is done by including a diode, D, in series with the resistor as in figure 3(c).

When the throttle is advanced to the required speed, the capacitor charges through the diode, which is biased in the forward direction and has little resistance. The starting inertia is therefore almost unaffected by the use of the diode.

When the throttle is returned to zero, however, the diode is biased in the reverse direction by the charge on the capacitor. It has a very high resistance and the time for the capacitor to discharge is considerably increased. This gives an unduly long "free-wheeling" distance for the train.

In figure 3(d), a variable resistor, Rb, has been added in parallel with the capacitor to give a more realistic free-wheeling distance. It has little effect on the starting inertia, but limits the free-wheeling distance when the throttle is returned to zero. The brake is "applied" by reducing the value of this resistance, which shortens the discharge time of the capacitor and gives a decreased stopping distance with the appearance of braking.

However, the brake resistor, when connected in this position, discharges the inertia capacitor slightly between each input pulse. It also forms a potential divider with the inertia resistor to further limit the voltage across the capacitor. This limits the maximum speed which can be obtained with a controller using this form of circuit.

In figure 3(e), the brake resistor Rb is placed in series with a diode Db and taken to the wiper of the throttle control instead of to the supply negative. While a charging current is flowing into the inertia capacitor, the voltage developed across the inertia resistor reverse biases the diode Db. The high impedance of this reverse-biased diode effectively isolates the brake from the inertia circuit.

When the voltage at the wiper of the throttle is less than that across the inertia capacitor (between successive peaks of the input supply voltage), the diode Db is forward biased. The inertia capacitor is then able to discharge through the brake resistor, which, however, is now in series with at least part of the throttle. This additional resistance reduces the discharge of the inertia capacitor between the supply peaks.

The reduced effect of the brake resistor on both the charge and discharge of the inertia capacitor results in a greater maximum speed. When the throttle is returned to zero, the brake is returned to the supply negative to give the same free-wheeling distance as in the circuit of figure 3(d).

The circuit of the new controller is shown in the main diagram. The thyristor (SCR) is a higher powered type, the G-E type C20D.

This circuit, like those of the earlier units, includes components for transient suppression, overload protection, and control. The suppression and protection are very similar to those employed in earlier circuits, while the control circuit is smaller to that developed in figure 3.

The diode D5 in the cathode circuit, and the capacitor C4 and the resistor R11 in the gate-cathode circuit of the thyristor are to remove transients which may be generated in the motor. D5 removes negative spikes, while other transients charge C4 which then discharges rapidly through R11. Transient suppression is necessary to prevent false firing of the thyristor, which would make speed regulation very erratic.

Protection against overload is provided by the resistor R13,the lamp, and the five-amp fuse. The resistor R13 limits the short surge current to less than 35 amps which is within the capabilities of the thyristor. The lamp takes a short time to warm up, but then limits the steady current in the track to under five amps which is less than the continuous rating of the thyristor. If a short circuit is held across the track, no damage can be done to the thyristor. The fuse limits the steady current through the thyristor to protect it against internal faults.

Working back from the gate of the thyristor we have T2, which is simply an emitter follower amplifying stage to ensure that there is adequate current for the thyristor gate. In the base circuit of T2 is the phase delay network, C3, R8, R9, with the emitter/collector circuit of T1 connected in parallel with R8, R9. An easy way to visualise the operation of T1 is as a variable resistor, its value at any time being determined by the current flowing into the base. By shunting R8 and R9 this resistance can be used to vary the delay of the phase delay network.

The base circuit of T1 is very similar to that in figure 3(e). variable speed control R2, or THROTTLE, has a pre-set resistor R1 in series to determine the maximum speed. The resistors R3 and R4 in series with the diode D1 give the simulated starting inertia. Resistor R3 is pre-set to adjust this inertia.

The resistor R4 limits the current through D1 in the worst possible case of peak supply voltage, fully discharged inertia capacitor, and minimum resistance in the throttle and inertia circuits.

The resistors R5 and R6 with the diode D2 between the throttle and the capacitor C2 form the braking circuit. The variable resistor R6, or BRAKE, has a pre-set resistor R5 in series to limit the braking distance according to whether realistic or emergency braking is required. A resistor R7 is included in series with the base lead of transistor T1 to ensure that the base current of the transistor cannot exceed a safe value, and also to limit the discharge of the inertia capacitor via the transistor between peaks of the supply.

On the assumption that several controllers will share a common power supply the RF choke L1 and the capacitor C1 have been included. If left out of the circuit, firing pulses from one control unit could be fed via the internal impedance of the power supply to trigger other controllers. If each controller has a separate power supply, these filter components may be omitted.

Before giving details of the construction of this controller, it is necessary to consider the choice of values for the components in the circuit and possible variations in those values to give maximum flexibility in design for the experimenter.

The inductance of the RF choke L1 is not critical and is similar to that used in both previous controllers. It consists of 50 to 60 turns of 24 or 26SWG enamelled wire on a short length of ferrite rod. Our rod was broken from an 8in length of ⅜in diameter ferrite made by Ducon.

When we modified the brake circuit as in figure 3(e), using a 1K throttle as in the previous controllers, there was only a moderate increase in maximum speed. Increasing the throttle to 10K, however, increased maximum speed quite noticeably. This difference in performance is because, with a total resistance for R5 and R6 of about 25K to 30K, the 10K for the throttle gives a substantial increase in total discharge resistance during the period between supply voltage peaks.

As the resistance of the throttle is made larger the maximum speed increases – although not in direct proportion – to a limit imposed by the supply voltage. On the other hand the extra throttle resistance increases the simulated inertia at intermediate throttle settings. This gives an unrealistic effect if the throttle is too high in value. Compromise values for this control lie in the range 5K to 20K.

If a power supply with several voltage tappings is used with a single controller, the maximum speed control R1 may not be necessary. However, if several controllers are used with a common supply, and/or if several different types of model loco are to be used with the layout, this control is essential. Its value is not critical, but the lowest value should be chosen which is consistent with safety for all locos likely to be used with the layout.

If several locos with different characteristics are used regularly with a controller, R1 could be replaced by a series of switched fixed resistors selected to give the required maximum speed for each individual loco. It is probably advisable, however, to retain a pre-set resistor as a further position on the switch if visiting locos are to be catered for.

The diode D1 in the inertia circuit can be of any type with a peak inverse rating of at least 40V.

The minimum possible inertia – with R3 set at zero – is determined by the value of R4, and may be increased by making R4 larger than the minimum necessary for the protection of D1.

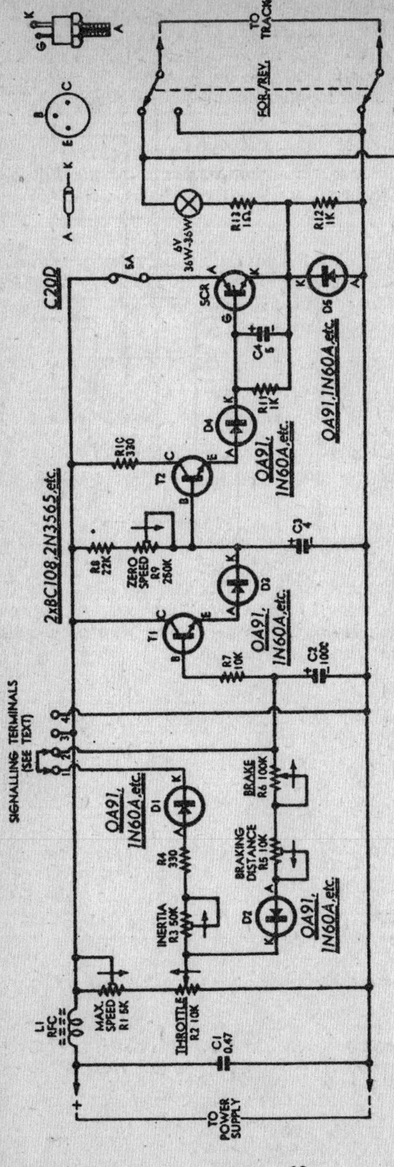

HIGH-POWER TRAIN CONTROLLER

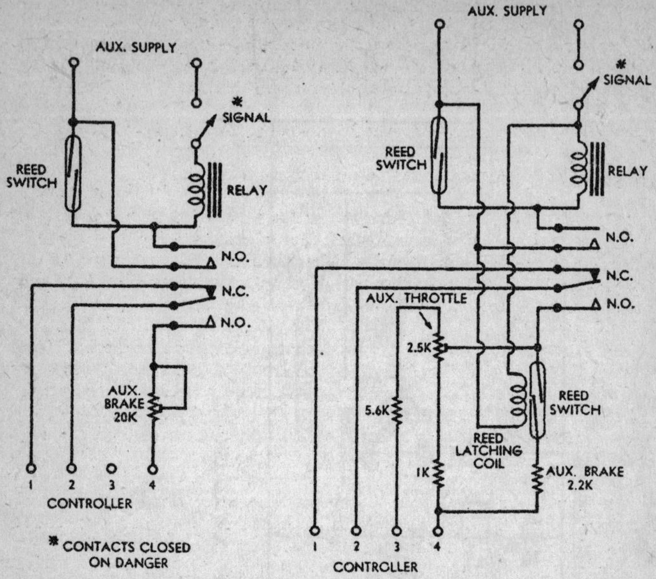

Figure 4. The signalling circuits with suggested modified values to suit this new controller.

The maximum inertia is determined by the value of the resistor R3. We found that values between 20K and 50K gave satisfactory results. If only low inertia is wanted, the resistance may be smaller. On the other hand, if very high inertia is required, R3 may be made 100K or more. However, as the resistance is increased, the maximum speed is reduced.

As in the case of the maximum speed control R1, the inertia control R3 (and R4) may be replaced by a series of switched fixed resistors to suit individual locos likely to be used with the controller.

For the brake control R5, we found that either a 100K linear or a 50K logarithmic potentiometer gave satisfactory results. A larger value – 200K lin or 100K log – gives a fractionally higher maximum speed, but the brake control loses its "feel" for the first part of its action. A smaller value for this control limits the top speed and should be avoided if possible.

An important factor restricting the maximum speed is the current drain from capacitor C2 through the base circuit of transistor T1. Resistor R7, originally fitted simply to limit T1 base current to a safe value, can be increased from its minimum value of 1K to restrict this current drain. We chose a value of 10K, although this could be increased to 22K with slight advantage in some circumstances. There is little point in any higher value.

All the foregoing improvements, i.e., the modified brake circuit and the increase of R2 and R7 to 10K, could be incorporated in last year's controller to give some increase in maximum speed.

The connection from the inertia resistance circuit – R3, R4 and D1 – to the inertia capacitor C2 is taken to external terminals for a signalling circuit of either type described in "Electronics Australia" January 1968. If the signalling facility is not required, the terminals should be connected by a link (shown in the circuit diagram.)

Because of the modifications to the inertia circuit in this controller compared with the 2A version, we found it necessary to alter some of the values in the signalling circuits. The new values we selected are shown in the accompanying circuits, figure 4.

We must stress that the values for these circuits depend on the supply voltage and on the particular loco being used. However, the order of change which we found necessary should give a starting point for experiments in modellers' own circuits. All our experiments with the controller and the signalling circuits have been conducted with an AC supply to the main rectifier of 15V.

In our prototype controller, transistors T1 and T2 are low voltage types (2N3565 or BC108) suitable only with supply voltages not greater than 20. If a higher voltage unit is required, transistors of higher voltage ratings should be used, such as the 2N3567 or BC107. If other transistor types are used, the maximum current rating should be about 100mA or higher, while the maximum voltage rating should be equal to or greater than 1.5 times the supply voltage.

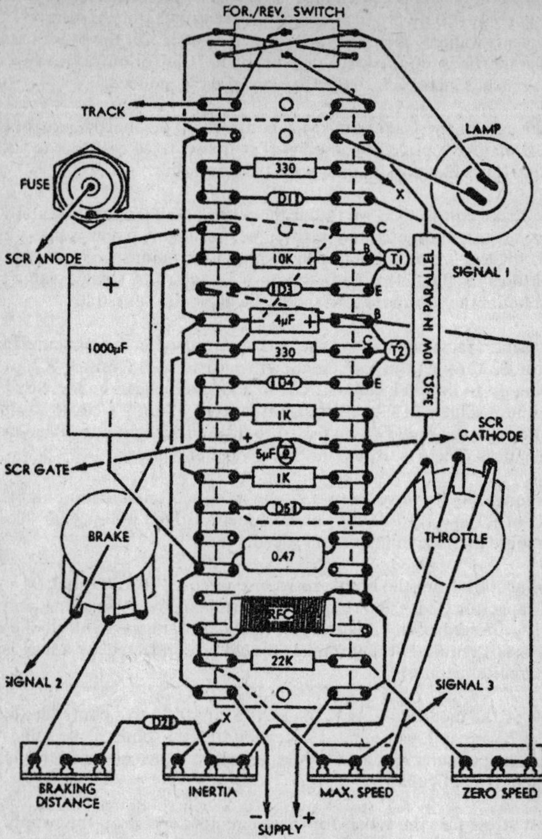

Complete wiring diagram showing the layout of components on the terminal panel and the interconnections to the major components. This is a rear view of the layout.

Diodes D3 and D4 are in the emitter circuits of transistors T1 and T2 respectively to protect the transistors against reverse base-emitter voltages. Diodes with reverse voltage ratings of 25V are adequate.

The main function of resistor R8 is to limit the base current of transistor T2 to a safe maximum, and for this needs a minimum value of 10K. However, we found that under some conditions of supply voltage and loco we needed a slightly greater resistance than the maximum value of the zero speed control R9. This is obtained conveniently by increasing R8.

Capacitor C4, for transient suppression, is connected in the cathode circuit of the thyristor in our previous train controllers. However, this controller has a tendency to oscillate at about 1MHz with the capacitor in this position. Moving it to the gate-cathode circuit and increasing its value overcomes this and still gives the required transient suppression.

The surge limiting resistor R13 should have a power rating of not less than 25W. The one in our prototype is made up of three 3-ohm resistors in parallel, each with a 10W rating.

Our model of the controller is constructed in a 6in x 6in x 6in sloping-front case from Heating Systems, with a 6in square front panel manufactured by that company for our 2A inertia controller – we changed the fuse designation to 5A. The pre-set controls are on a bracket in the case below the front panel for easy access. The size and layout are not critical.

The thyristor is mounted on a 5½in x 2½in x 16 gauge aluminium heat sink as a precautionary measure. This is supported on insulated stand-offs at the rear of the case.

The connections to the power supply, to the track, and to the signalling circuits are taken via terminals on the back of the case. Only three signalling terminals are shown, the negative "supply" terminal being used as terminal "4". However, an actual terminal may be fitted if preferred, simply wired directly to the negative supply terminal.

The brake potentiometer is mounted on a bracket to allow a spring to be fitted behind the panel. This spring gives the brake a mechanical resistance and makes it feel more realistic. We tried a number of springs, and finally settled for a spiral type normally used in simple domestic toasters. Several types are available, but we chose the heaviest we could find. This worked out just right. The throttle potentiometer is also mounted away from the front panel to cater for the handle. (A generator handle from an ex-army field telephone.)

The wiring diagram shows the arrangement of the components on the panel. Those on the component board should be mounted before it is fixed to the panel. The overload lamp is wired directly into the circuit using 16 or 18SWG tinned copper wire to support the lamp firmly behind the bezel.

When soldering the components, particularly the semiconductors, care must be taken not to damage them. Leads should be bent no nearer than 3/16th of an inch from the body, and they must be held in wiring pliers while being soldered. The soldering iron should be as hot as possible to ensure that the joint is made quickly.

The pre-set controls are arranged on the front panel so that they are normally adjusted from left to right. Before starting to adjust the controls, set all of them fully anti-clockwise. As all model-train locos have their own characteristics, it is advisable to check each setting with the various trains to be used. A closed loop of track is best for the adjustment procedure.

The first adjustment is to set the zero speed control (ZERO SP) so that all locos remain stationary with power applied and the throttle at minimum. Place a loco on the closed loop, making sure the throttle is set at zero. When the power is applied the engine will move around the loop. Turn the zero speed control clockwise until the engine stops and advance the control about 30 degrees further. Try all other locos to ensure that none of them moves. Should any move, readjust the control.

The maximum speed control (MAX SP) is set so that with the throttle at maximum, no loco is going so fast as to be in danger of derailing. If it is travelling too fast, reduce the supply voltage. If it is too slow, advance the pre-set control.

Note particularly: The supply voltage should be no higher than is necessary to give the required top speed. Excessive supply voltage can cause jerky operation at low speeds, due to the thyristor being so close to the non-firing condition that it may not fire on all cycles.

When some inertia is introduced, the top speed is reduced to give a safety margin for the locos. With the fastest loco on the loop, rotate the throttle to maximum, and time the train over a measured distance.

The inertia pre-set control (INERTIA) is set according to personal requirements, as a wide range of acceleration is possible. The fastest rate allows an immediate start at the speed set by the throttle. The slowest rate allows a pause of several seconds before the train starts to move even when the throttle is wide open. This is followed by a slow acceleration to the speed determined by the throttle. The braking distance pre-set control (BR DIST) is adjusted on a similar basis. A train can be stopped instantaneously if desired, or it can take up to about eight (actual) feet to stop a train from maximum speed. When making this adjustment, we reduced the time taken to reach maximum speed by placing a short circuit across the inertia control (with a screwdriver blade) for a second or two. This accelerates the loco very quickly to its maximum speed.

PARTS LIST

1 Sloping front case, 6in x 6in x 6 in.
1 Front panel, 6in x 6in.
1 Heat sink (for thyristor), 5½in x 2½in x 16g aluminium.
1 Tension spring (see text).
1 Fuse holder.
1 5-amp fuse.
1 Large bezel assembly.
1 6-volt 36-36watt twin filament lamp.
1 RF choke (see text).
1 20-pair length of miniature resistor panel.
2 Control knobs or handles, as required.
8 Terminals. Assorted colours.

SEMICONDUCTORS
1 Thyristor, type C20D.
2 Transistors, type BC108, 2N3565, RS276-2009.
5 Diodes, type OA91, 1N60A or RS276-1102/1136

CAPACITORS
1 0.47uF, 160V working.
1 4uF, 40V working.
1 5uf, 15V working (tantalum).
1 1000uF, 50V working.

RESISTORS
3 3-ohms, 10 watt.
2 330 ohms, ½ watt.
2 1K, ½ watt.
1 5K pre-set pot.
1 10K, ½ watt.
1 10K pre-set pot.
1 10K linear pot.
1 22K, ½ watt.
1 50K pre-set pot.
1 100K linear pot.
1 250K pre-set pot.

MISCELLANEOUS
Nuts, bolts, washers, solder tags, solder, wire, insulating sleeving, grommets, rubber feet, aluminium (16g) for brackets and heat sink, etc.

AN ELECTRONIC STEAM WHISTLE

Our "Electronic Steam Whistle" was developed with the model railway enthusiast in mind. It might also serve as a sound effects simulator for amateur theatricals. Featuring single and dual horns, with background steam, it will add realism to your model setup. A separate 'Noise" outlet is also provided to simulate steam alone.

We considered it essential that all the factors affecting the final sound should be capable of being varied over a wide range; preferably a wider range than appeared to be strictly necessary. This would permit the builder to vary any or all of these factors, as he thought fit, in order to achieve what he considered the best imitation of the particular whistle he had in mind.

If we consider the type of sound a steam whistle makes, we will be in a better position to decide how we can imitate it electronically. For a start, when the steam valve is opened, the sound intensity appears to rise to a constant level over a short period. In other words, it has a certain rise time. Also, during this rise time, the frequency of the whistle falls slightly.

Now consider the effect as the steam valve is closed. The steam in the whistle will take a short period to escape (as the pressure drops), thus the sound intensity fades over a short period. In other words the whistle has a certain decay time. Another feature peculiar to whistles is the background sound of the steam or air which operates them. As mentioned earlier, our device also features dual horns. This type of sound is usually produced by air horns. The two horns having a frequency ratio of about 1.5 to 1.

We must now design circuitry which will imitate the above features. To avoid confusion, we shall forget the dual horns for the present. To produce the single whistle, we require an oscillator. A simple phase shift oscillator would appear to be suitable. (Tr1 and associated components). There are no inductances required, and the frequency can be varied easily by changing resistance or capacitance.

To simulate the steam, a logical choice is a white noise generator. This is merely a reverse biased diode, arranged so that avalanche takes place, producing a substantial amount of noise. We use a reverse biased BC108 (D3) emitter to base junction in this mode.

Initially, we tried mixing the output from the noise generator directly with the oscillator output, but we were not happy about the level of noise available. It may have been sufficient in some cases, but it was a marginal condition. Accordingly, we fitted the noise amplifier (TR3) which provides more than sufficient noise for any likely requirement.

The signals from the oscillator and noise generator must be mixed in the correct ratio before they are fed to the main amplifier, and they must only reach the input to this amplifier when we wish to initaite the whistle. Also a means to vary the rise and decay times of the signal must be devised. All these functions are conveniently performed in the gated amplifier stage (TR4).

The whistle and noise signals are mixed at the input to the gated amplifier. The ratio of these signals is adjusted by resistance in series with the coupling capacitors from these stages.

The gated amplifier is normally gated off by a voltage divider network consisting of the 4.7K in the emitter circuit and the 47K to the positive rail. To gate the amplifier on, in the simplest case, it would be sufficient to shunt the emitter resistor with another resistor of suitable value. However, we can provide the required attack and decay times by adding suitable time constant circuits which control the rate at which the stage is gated on or off.

The time constant circuits consist of the 1K and 2.2K resistors and the two 10uF electrolytic capacitors in the emitter circuit of TR4. The total resistance of 3.2K is that required to gate the amplifier on, but the rate at which this can happen is determined by the rate at which the capacitors can be charged or discharged through their associated resistors. The 10uF capacitor across the 4.7K emitter resistor also functions as a conventional by-pass to maintain the amplifier gain.

When the test button is pressed, the lower leg of the voltage divider to TR4 emitter becomes approximately 1.9K ohms. However, before the stage is biased on, the emitter bypass capacitor must discharge through the emitter resistor in parallel with the gating resistance, and the other electrolytic must discharge via the 1K resistor.
Thus the output of the gated amplifier rises to a constant level over a short period, as the electrolytics discharge. We have thus introduced the necessary rise time. This rise time can be increased by increasing the capacitance of either electrolytic, or the gating resistance, and vice versa for a decrease. Note: If the gating resistance is increased too much, the stage may not bias on at all, thus it is preferable to vary only the capacitance if it is desired to change the rise time.

These electrolytics also provide the decay time. When the test button is released, the emitter bypass capacitor charges via the 47K, while the other electrolytic charges through the 2.2K before the gated amplifier is biased off. Thus the output from this amplifier fades over a short period. The rise and decay time adjustments interact, thus a compromise must be accepted, but more about this later.

Earlier we mentioned that a steam whistle appears to lower pitch during the rise time. The method we used to achieve this effect is as follows: The gated amplifier output signal is fed via a 0.22uF capacitor to a rectifier (D2). The DC thus produced is proportional to the output signal amplitude and, therefore, increases during the rise time. However, C2 (DC reservoir) must charge before a constant negative potential exists, thus the value of C2 controls the rate at which this potential increases.

In OSC1 a BA100 (D1) is connected in series with an 82K from TR1 base. Normally, D1 is biased on, thus the 0.0047 is shunted by 82K. Under these conditions the oscillator frequency is approximately 780Hz. If the DC control voltage is now applied to the anode of D1 via a 2.2k, this diode will be biased off. The effect of the 82K as a discharge path now becomes negligible, thus the overall phase shift of the network increases, and the frequency settles at approximately 750Hz.

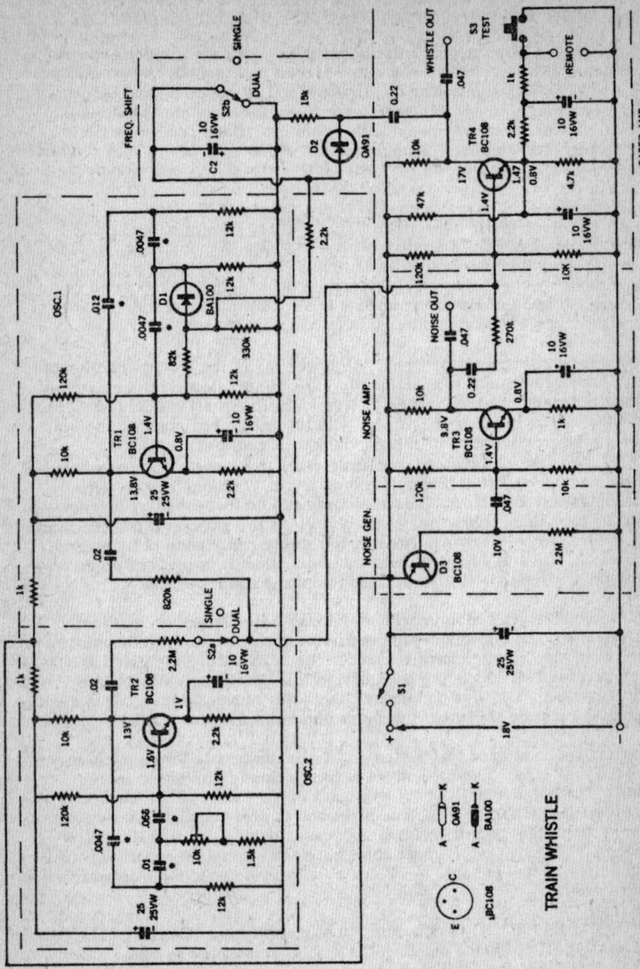
OSC.1
OSC.2
FREQ. SHIFT
NOISE GEN.
NOISE AMP.
GATED AMP.
WHISTLE OUT
NOISE OUT
REMOTE
S3 TEST
TRAIN WHISTLE

OSC2 was added to provide the dual horn feature. In this mode, SW2b disables the DC control voltage, thus OSC1 operates at approximately 750Hz continuously. Meantime, SW2a switches the 2.2M in series with the OSC2 coupling capacitor to the gated amplifier base. Now when the test button is pressed, we hear the dual tone plus noise. The rise and decay effects still occur, but there is no frequency shift. The final frequency of our second oscillator was found to be approximately 490Hz for the most realistic dual sound.

The 10k preset pot in one leg of the OSC2 phase shift network was included to allow the frequency ratio between oscillators to be finely adjusted, but more about this when we discuss final testing.

Construction should not present any serious problems for the average reader with some previous experience. While the finished product may look a little complicated, it must be remembered that it is a collection of separate sections. To make things easier, we have described the construction on a section by section basis, with details of how to test each section before preceeding to the next one.

We constructed the two oscillators on a piece of tag board, 17 pairs of terminals long. The noise generator, noise amplifier, frequency shift network and gated amplifier are built on a second length, 19 pairs long. We suggest the oscillator board be wired first but, if the dual horn facility is not required immediately, the OSC2 section may be omitted for the present. To allow the tag boards to fit into the small case we chose, it was necessary to limit the number of tags on each board. This made it necessary to place some components on the underside of each board.

Two threaded pillars hold the tag boards apart, while another two secure the whole assembly in the case. We cut the heads off two 1/8 Whit. screws and used these as studs to sandwich the oscillator board between the pillars. Take care that no components are placed where the pillars are to be located.

If one follows the circuit and wiring diagrams, wiring OSC1 should be relatively straightforward. There are only two components under the board the 82K and the 25uF electrolytic. Where a capacitor or resistor is soldered to tags which have at least one other tag between them, the pigtails of that component should be insulated with spaghetti, to eliminate any possible shorts. Also insulate the pigtails of those components under the board. With this section wired, check it against the OSC1 section of the circuit, and the wiring diagram.

At this stage, OSC1 can, and should, be tested. Testing will call for an amplifier or, better still, an oscilloscope. As a last resort a high impedance headphone will suffice. Connect an 18V positive battery terminal to the free end of the 1K resistor and the negative terminal to the negative rail. Connect the 820K resistor to the amplifier, oscilloscope, or headphone, either of which should indicate a sine wave output. It must be realised however, that much of the realism is still missing. Do not make a final judgment until the noise (steam), attack and decay, and frequency shift have been added.

OSC2 occupies the rest of this board. The same procedure as that described for OSC1 applies to this section. The only extra component is the 10K preset pot, although the phase shift capacitor values have been changed to lower the frequency. One of these capacitors (.068uF) is situated under the board, along with the 25uF electrolytic, and a 1.5K and a 2.2K resistor.

To complete this board, interconnect the oscillator negative rails with a short length of wire as shown on the wiring diagram.

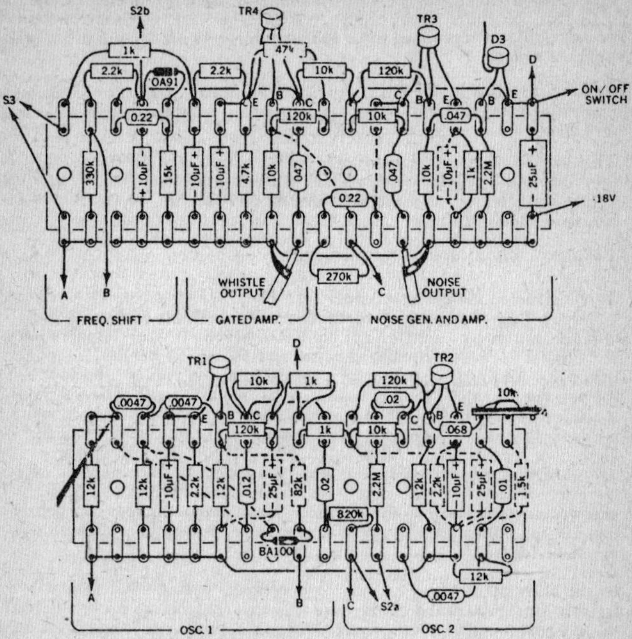

Wiring diagrams of the two boards. The function of each section is clearly shown.

OSC2 can be tested in the same manner as OSC1, but Rv1 should be varied during the test, to ensure that the pitch of OSC2 does vary noticeably with this adjustment.

To begin the second board we can wire the noise generator. This section has only three components, so it should not take long to complete it. The collector of the BC108 is not used, so simply bend this lead up out of the way.

The noise amplifier is logically situated next to the noise generator. The 10uF emitter bypass is the only component under the board. The positive rail for this section is via a wire to the emitter of the noise generator transistor. Once again the sections should be checked against the circuit before testing.

To test the noise generator and amplifier, connect the positive 18V terminal to the emitter of the noise generator transistor (used as a diode) and the negative terminal to the negative rail. Connect the free end of the .047uF capacitor (labelled Noise Output) to the input of your amplifier or oscilloscope. If there is no output, place a fairly large value capacitor in series with the amplifier input lead and check if there is output from the noise generator at the amplifier base.

The rest of this board accommodates the gated amplifier and frequency shift network. The 1K gating resistor mentioned earlier is the only component under the board. Make sure that the negative end of C1 (refer to frequency shift network) connects to the OA91 anode. A short length of wire from the gated amp collector to the 0.22uF capacitor (shown on the wiring diagram), and another length to connect the negative rail, completes this board. Check the complete section against the circuit and wiring diagram.

The gated amplifier can be tested using the amplifier or oscilloscope, as for earlier tests. Until the test button is pressed, there should be no output from the gated amplifier, although there will be signal at the base of TR4 continuously. The emitter voltage under these conditions should be approximately 1.47V. When the test button is pressed, there should be output, and the emitter voltage should be approximately 0.8V.

If the output is observed on a CRO, there should be some distortion obvious, one half cycle being more rounded than the other. This is deliberate, and adds a little to the realism of the sound. The background noise (steam) should also be obvious as "grass" superimposed on the waveform. If only an amplifier is available, simply check that the stage gates on and off correctly, and that the level of noise is adequate.

The rise and decay of the output signal is relatively fast, so if there is need to prove that these effects are present, remove the two electolytics from the emitter circuit of TR4. If the gate is now opened there should be a noticable difference in the attack and decay characteristic.

Four wires interconnect the two boards. Make these long enough to allow future access to the boards should it be necessary. If the two boards are fastened one above the other (with the oscillator board on the bottom) using the threaded pillars, the assembly should sit rigidly while the flying leads are soldered. Make these leads long enough to allow the boards to be removed from the case while the unit is operating, to allow for service or adjustment. Connection from the boards to the whistle and noise outlet sockets is by shielded cable.

While we have mounted the complete unit in a metal box and some readers may elect to do the same, this is not essential. Other readers may prefer to wire the boards directly into their existing model railway electrical system, with leads running directly to the various ancillary devices.

Those who make their own boxes should be particularly careful in positioning the mounting holes for the boards. Any serious error here may locate the terminals on the boards too close to the metal side of the box, with consequent risk of short-circuits.

The train whistle should now be complete, apart from final testing and adjustment. As in the case of the initial tests, an 18V battery is required, conveniently made up from two nine volt batteries as specified ind the parts list. Output from the whistle is approximately 500mV PP across 50K ohms, so the amplifier it is to drive need not be particularly sensitive. If the amplifier is very senstive, it would be advisable to place a 50K preset pot across the whistle output, and adjust the output to a level which does not overload the input stage of the amplifier. The noise output socket should also give approximately 500mV across a 50K ohm load.

Connect the whistle output to the input of the amplifier via a suitable lead. Switch to single horn, and press the test button. The whistle pitch should fall as this putton is pressed. The rise and decay effects may not be so noticeable, as they are much faster, but if they are eliminated the difference will be obvious. The whistle should be accompanied by background "steam".

If the whistle to steam ratio is not suitable, the 820K (whistle) or the 270K (noise) resistors connected to TR1 base can be changed. More resistance for less noise or whistle and vice versa. The oscillator pitch can be vaired by changing any of the capacitors in the OSC1 phase shift network. These are marked with asterisks. More capacitance lowers the pitch and vice versa. If the pitch or whistle to noise ratio is altered it must be appreciated that OSC2 pitch or output must also be altered.

If C2 is increased, the rate at which OSC1 changes pitch will be decreased. If the 2.2K connecting to C2 is increased, the amount by which it changes pitch will be decreased.

As mentioned earlier, the rise and decay time settings interact to some extent. If both the electrolytic capacitors in the TR4 emitter circuit are increased in value by the same amount, the rise and decay times will increase together, and the ratio between these times will remain relatively constant. Likewise if both capacitors are lowered in value the rise time and the decay times will decrease.

If the reader desires to vary the ratio between rise and decay times, the TR4 emitter bypass electrolytic will mainly affect decay time. More capacitance for more decay time and vice versa. Likewase the 10uR electrolytic connected to the junction of the 1K and the 2.2K will mainly affect the rise time.

If OSC2 has been included, switch to dual horn and short the remote socket so that the whistle is on constantly. The frequency of OSC2 can be varied broadly by changing the capacitors marked with asterisks, as for OSC1, but this should only be necessary if OSC1 pitch has been altered. For fine frequency adjustment use Rv1. The dual horn sound can now be adjusted to suit your impression of how it should sound by adjusting the frequency of OSC2. Normally, this frequency will be correct when there is minimum low frequency beat discernible in the output. The output of OSC2 can be varied by changing the 2.2M in series with the .022uF coupler, but again this should only be necessary if OSC1 output has been changed.

The unit may be triggered remotely by any suitable means, using the remote socket. It could be triggered by placing a reed switch between the rails, and a permanent magnet under the loco. This arrangement would require a suitable holding circuit to maintain the whistle for a short period after the pulse occurs.

PARTS LIST

1 Metal box $5\frac{1}{4}$" (135mm) x 3 1/8" (79mm) x 2 1/8" (54mm)
2 Belling Lee chassis sockets, L604/S
2 Belling Lee plugs L734/P
1 2 pin speaker socket and plug to match
1 Miniature toggle switch
1 Miniature toggle switch, 2 pole 2 way
1 Miniature push button switch, normally off.
1 length miniature tag board, 17 pairs of tags
1 length miniature tag board, 19 pairs of tags
2 9v batteries
2 Clip connectors to suit above.

RESISTORS ($\frac{1}{2}$ watt)

2 2.2M	1 15K
1 820K	6 12K
1 330K	6 10K
1 150K	1 4.7K
1 270K	4 2.2K
4 120K	1 1.5K
1 82K	4 1K
1 47K	

1 10K preset miniature pot.

CAPACITORS

Electrolytics
3 25uF 25VW, 6 10uF 16VW
Miniature Polyesters (100VW)
2 0.22uF, 3.047uF, 2.022uF, 1 .01uF,
3 .0047uF
Polyesters (160VW)
1.068uF, 1 .012uF

SEMICONDUCTORS

5 BC108 transistors, or equivalent
1 BA100 silicon diode
1 OA91 diode

A MODEL TRAIN CHUFFER

This novel sound effect simulator will add realism to model railway setups, by simulating the familiar steam train "chuff chuff" sound. The chuff rate is variable via an external control, allowing the user to synchronise the sound to the speed of the model.

Our chuffer was developed as follow up to the Steam Whistle. We were not aware of any similar type of sound effects simulator, thus it was necessary develop this unit from scratch.

Most readers will be familiar with the sounds a steam loco makes as it stands in a station. These vary, but in many cases there is a low level hiss of steam punctuated by chuffs from the Westinghouse braking system. Our chuffer can simulate this type of sound when the chuff rate control is set at minimum.

As the loco moves out of the station, the low level hiss disappears and the chuffs (now from the exhaust valves) become more frequent. At this time the duration of each chuff is short compared to the time between chuffs.

Eventually the loco reaches a constant speed and, in most cases, the chuff duration now appears to equal the time between chuffs. When the driver decides to slow down to reduces or shuts off steam to the cylinders. Thus the chuff sound may cease completely, or continue at a lower level and a decreasing rate. Appropriate manipulation can produce these effects.

The steam component for our chuffer is derived from the white noise output on the steam whistle as described previously.

The chuffer circuit consists of a diode gate and a modified sinewave multivibrator. The diode is merely a reverse biased diode which is biased on by one half cycle of the oscillator output. The output from a multivibrator can be made roughly sinusoidal by introducing certain amounts of positive and negative feedback. Our oscillator is designed so that the output waveshape changes with frequency.

At minimum chuff rate there is a short positive pulse, after which the output goes negative and then decays exponentially to zero, where it remains for several seconds. During this off time the diode gate is reverse biased, and only the most positive peaks of the white noise pass through it. This results in a faint background hiss between chuffs which, while somewhat accidental, is not out of keeping with the sound normally encountered in the vicinity of a loco.

When the positive pulse recurs, the diode is forward biased to allow the maximum noise through. At minimum chuff rate our unit gave one chuff approximately every 2.5 seconds, or 24 chuffs per minute.

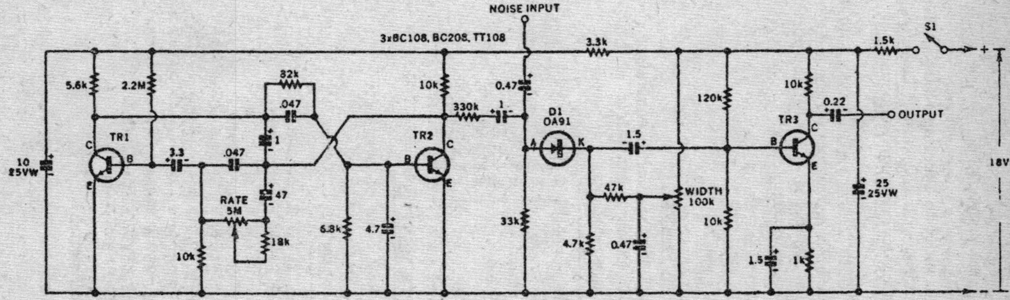

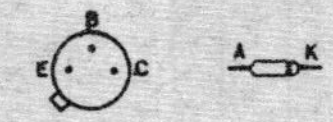

CHUFFER

The circuit is a relatively simple one. Transistors TR1 and TR2 form the multivibrator which controls the chuff rate. The rate is controlled by the 5M pot. Output from the multivibrator gates the diode D1 on and off, allowing the white noise from the "noise, input" terminal to pass through it in short bursts. Bias applied to D1 controls the period it is open and, hence, the chuff width.

As the chuff rate is increased, the positive excursion of the oscillator output, becomes roughly sinusoidal, while the duration of the negative excursion decreases until at maximum chuff rate, the positive and negative excursions are approximately equal. The maximum chuff rate from our unit was approximately 12 per second.

The oscillator frequency is varied by shunting a 0.047uF feedback capacitor with a 47uF in series with a 5M plot. (Base circuit of TR1). Minimum resistance gives minimum frequency.

The pot marked "Chuff Width" allows variation of the diode reverse bias, thus the diode can be made to conduct for the full duration of the positive excursion of the oscillator output, or for any portion of it. Maximum bias cuts the diode off completely.

PARTS LIST

SEMICONDUCTORS

3 x BC108 Transistors or equivalent
1 x OA91 Diode or equivalent

RESISTORS (½ watt)

1 x 5. 6K	1 x 33 K
4 x 10 K	1 x 3. 3 K
1 x 2. 2 M	1 x 47 K
1 x 18 K	1 x 4. 7 K
1 x 82 K	1 x 120 K
1 x 6. 8 K	1 x 1 K
1 x 330 K	1 x 1. 5 K

POTENTIOMETERS

1 x 100 K
1 x 5 M

CAPACITORS

1 x 10uF 25VW Electrolytic
1 x 3. 3uF 25VW Electrolytic
2 x 1uF 25VW Electrolytic
1 x 47uF 25VW Electrolytic
1 x 4. 7uF 25VW Electrolytic
2 x 0. 47uF 25VW Electrolytic
2 x 1. 5uF 25VW Electrolytic
1 x 25uF 25VW Electrolytic
1 x . 22uF 25VW Electrolytic
2 x 0. 047uF Polyester

1 x On-off switch

EQUIVALENT SEMICONDUCTORS

TT801	BFR40, RS276-2018, 2N4921, BD131, TIP29, 2SC496
2N3055	BD130, BDX10, BPY20-39, 2N3713, RS276-2020, AM219, SK3027, CV8889.
EM401	OA626/50, BY126/50, SD55-05, RS276-1139-1101-1105, BY127, 1N4002-4007, BAY219.
OA91	AA117-118-132-144, OA81-95-161, SK38, SFD108, 1N38-55-63-617, 1S33, RS276-1102/1136.
BC108	2N3565, TT108, TT3565, BC107, RS276-2009, AM252.
BA100	BA108-127-128-147-188-216-218-219, BAX16, BAY18, OA129, 1N457-914-4446-4447-4448-4449, 1S72-181, 52J2.

N. B. Although equivalent semiconductors may have similar electrical properties, physical dimensions may be different and this must be born in mind if space is tight and for mounting details.
Remember polarities when replacing PNP with NPN and vice versa. If in doubt always be advised by your local dealer for suitable equivalent semiconductors.

Details of a selection of Varta Nickel-Cadmium Batteries and cells that may be of use in the previous chapter on lighting.

BUTTON CELLS (Average discharge voltages approx. 1.2V; charging voltages approx. 1.4V - 1.5V)

Type No.	Capacity 10 Hrs. rate	Weight (Grms)	Dimensions (mm)		
			Dia.	Height	
10DK	10mAh	0.9	7.6	5.0	
20DK	20mAh	1.1	11.4	5.1	
50DK	50mAh	3.5	15.5	5.85	
150DK	150mAh	11	25	6.6	
225DK	225mAh	12.5	25	8.6	
450DK	450mAh	33	43	7.6	
1000DK	1mAh	57	50.3	10	
225DKZ	225mAh	13	25	8.8	
500DKZ	500mAh	26	34.3	9.5	
CYLINDRICAL CELLS					
151 D	150mAh	12	12	29	
900 D	900mAh	40	13.5	90	
BUTTON CELL BATTERIES					Nominal Voltage
5/150DK	150mAh	55	26	34.5	6V
10/150DK	150mAh	115	26	68	12V
5/225DK	225mAh	65	26	46	6V
10/225DK	225mAh	135	26	91	12V

Varta (Great Britain) Limited
Hermitage Street
Crewkerne
Somerset
TA18 8EY

Details of a selection of SAFT Nickel-Cadmium batteries and cells that may be of use in the previous chapter on lighting.

BUTTON CELLS (Average discharge voltage approx. 1.2V)

Type No.	Capacity 5 Hr Rate	Weight (Grms)	Dimensions (mm)	
			Dia.	Height
VB4	40mAh	3.5	15.7	6.0
VB10/VB10E	100mAh	7	23.0	5.3
VB22/VB22E	220mAh	12	25.1	7.8
VB30/VB30E	300mAh	18	34.7	5.5
VB60/VB60E	600mAh	31	34.7	9.8

CYLINDRICAL CELLS

VRO.11/3AA	0.100Ah	11	14.5	17.4
VRO.45 1/2A	0.450Ah	21	17.3	28.1
VRO.5AA	0.500Ah	24	14.5	50.2
VRO.7 1/2C	0.700Ah	39	25.9	26.2
VR1.2RR	1.2Ah	51	22.7	42.1

SAFT (United Kingdon) Ltd
Castle Works
Station Road
Hampton
Middx
TW12 2BY

Details of a selection of Ever Ready Nickel-Cadmium Batteries and cells that may be of use in the previous chapter on lighting.

BUTTON CELLS (Average discharge voltage approx. 1.2V)

Type No.	Capacity	Weight (Grms)	Dimensions (mm)	
			Dia.	Height
NCB22M	225mAh	13	25.0	8.8
NCB50M	500mAh	26	34.3	9.5
NCB100M	1000mAh	57	50.3	10.0

CYLINDRICAL CELLS

$\frac{1}{3}$ AA	0.12Ah	8	14.3	17.1
$\frac{1}{2}$ AA	0.24Ah	14	14.3	28.1
AA	0.50Ah	25	14.3	50.0
Super AA	0.60Ah	30	15.6	50.0
RR	1.20Ah	50	22.8	42.7

Ever Ready (Special Batteries) Ltd
Hockley
Essex
SS5 4AH

Details of a selection of Ever Ready bulbs that may be of use in the previous chapter on lighting.

Type of Bulb	Type No.	Voltage	Current	Description
Torch	1427	1.25V	0.25A	Round
Torch	2111	2.5V	0.3A	Spotlight
Torch	2112	3.5V	0.3A	Spotlight
Torch	2114	5.5V	0.3A	Spotlight
Torch	2113	1.2V	0.25A	Lens End
Torch	2225	2.2V	0.25A	Lens End
Torch	2170	2.5V	0.3A	Pre-focus
Torch	2171	3.8V	0.3A	Pre-focus
Torch	2172	3.6V	0.5A	Pre-focus
Torch	2174	4.75V	0.5A	Pre-focus
Torch	2175	2.4V	0.5A	Pre-focus
Torch	2176	5.95V	0.5A	Pre-focus
Krypton	2182	3.6V	0.5A	Pre-focus
Krypton	2183	4.75V	0.5A	Pre-focus
Krypton	2184	5.4V	0.54A	Pre-focus
Flasher	2573	4.5V	0.3A	Round
Dynamo	2625	6.0V	0.25A	Round
Dynamo	2630	6.0V	0.3A	Round
Dynamo	2650	6.0V	0.5A	Round
Dynamo	2610	6.0V	0.1A	Round
Panel	2272	6.5V	0.3A	Round

Ever Ready (Great Britain) Ltd
Ever Ready House
1255 High Road
Whetstone
London N20 OEJ

NOTES

NOTES

NOTES

Please note overleaf is a list of other titles that are available in our range of Radio and Electronics Books.

These should be available from most good Booksellers, Radio Component Dealers and Mail Order Companies.

However, should you experience difficulty in obtaining any title in your area, then please write directly to the publishers enclosing payment to cover the cost of the book plus adequate postage.

If you would like a copy of our latest catalogue of Radio & Electronics Books then please send a Stamped Addressed Envelope to:–

BERNARD BABANI (publishing) LTD
THE GRAMPIANS
SHEPHERDS BUSH ROAD
LONDON W6 7NF
ENGLAND

BP1	First Book of Transistor Equivalents and Substitutes	60p
BP2	Handbook of Radio, TV and Ind. & Transmitting Tube & Valve Equiv.	60p
BP6	Engineers and Machinists Reference Tables	50p
BP7	Radio and Electronic Colour Codes and Data Chart	25p
BP11	Practical Transistor Novelty Circuits	40p
BP14	Second Book of Transistor Equivalents	1.10p
BP22	79 Electronic Novelty Circuits	75p
BP23	First Book of Practical Electronic Projects	75p
BP24	52 Projects using IC741	95p
BP25	How to Build Your Own Electronic and Quartz Controlled Watches & Clocks	85p
BP26	Radio Antenna Handbook for Long Distance Reception & Transmission	85p
BP27	Giant Chart of Radio Electronic Semiconductor & Logic Symbols	60p
BP28	Resistor Selection Handbook (International Edition)	60p
BP29	Major Solid State Audio Hi-Fi Construction Projects	85p
BP30	Two Transistor Electronic Projects	85p
BP31	Practical Electrical Re-wiring & Repairs	85p
BP32	How to Build Your Own Metal and Treasure Locators	1.00p
BP33	Electronic Calculator Users Handbook	95p
BP34	Practical Repair & Renovation of Colour TV's	95p
BP35	Handbook of IC Audio Preamplifier & Power Amplifier Construction	1.25p
BP36	50 Circuits Using Germanium, Silicon and Zener Diodes	75p
BP37	50 Projects Using Relays, SCR's and TRIAC's	1.10p
BP38	Fun & Games with your Electronic Calculator	75p
BP39	50 (FET) Field Effect Transistor Projects	1.25p
BP40	Digital IC Equivalents and Pin Connections	2.50p
BP41	Linear IC Equivalents and Pin Connections	2.75p
BP42	50 Simple L.E.D. Circuits	75p
BP43	How to make Walkie-Talkies	1.25p
BP44	IC 555 Projects	1.45p
BP45	Projects on Opto-Electronics	1.25p
BP46	Radio Circuits using IC's	1.35p
BP47	Mobile Discotheque Handbook	1.35p
BP48	Electronic Projects for Beginners	1.35p
BP49	Popular Electronic Projects	1.45p
BP50	IC LM3900 Projects	1.35p
BP51	Electronic Music and Creative Tape Recording	1.25p
BP52	Long Distance Television Reception (TV–DX) for the Enthusiast	1.45p
BP53	Practical Electronic Calculations and Formulae	2.25p
BP54	Your Electronic Calculator and Your Money	1.35p
BP55	Radio Stations Guide	1.45p
BP56	Electronic Security Devices	1.45p
BP57	How to Build your own Solid State Oscilloscope	1.50p
BP58	50 Circuits using 7400 Series IC's	1.35p
BP59	Second Book of CMOS IC Projects	1.50p
BP60	Practical Construction of Pre-Amps, Tone Controls, Filters and Attenuators	1.45p
BP61	Beginners Guide to Digital Techniques	95p
100	A Comprehensive Radio Valve Guide – Book 1	40p
121	A Comprehensive Radio Valve Guide – Book 2	40p
126	Boys Book of Crystal Sets	25p
138	How to Make Aerials for TV (Band 1–2–3)	25p
143	A Comprehensive Radio Valve Guide – Book 3	40p
157	A Comprehensive Radio Valve Guide – Book 4	40p
160	Coil Design and Construction Manual	75p
178	A Comprehensive Radio Valve Guide – Book 5	40p
196	AF–RF Reactance – Frequency Chart for Constructors	15p
200	Handbook of Practical Electronic Musical Novelties	50p
201	Practical Transistorised Novelties for Hi-Fi Enthusiasts	35p
202	Handbook of Integrated Circuits (IC's) Equivalents and Substitutes	1.00p
203	IC's and Transistor Gadgets Construction Handbook	60p
205	First Book of Hi-Fi Loudspeaker Enclosures	75p
206	Practical Transistor Circuits for Modern Test Equipment	60p
207	Practical Electronic Science Projects	75p
208	Practical Stereo and Quadrophony Handbook	75p
210	The Complete Car Radio Manual	1.00p
211	First Book of Diode Characteristics Equivalents and Substitutes	95p
213	Electronic Circuits for Model Railways	1.00p
214	Audio Enthusiasts Handbook	85p
215	Shortwave Circuits and Gear for Experimenters and Radio Hams	85p
216	Electronic Gadgets and Games	1.00p
217	Solid State Power Supply Handbook	85p
218	Build Your Own Electronic Experimenters Laboratory	85p
219	Solid State Novelty Projects	85p
220	Build Your Own Solid State Hi-Fi and Audio Accessories	85p
221	28 Tested Transistor Projects	95p
222	Solid State Short Wave Receivers for Beginners	95p
223	50 Projects using IC CA3130	95p
224	50 CMOS IC Projects	95p
225	A Practical Introduction to Digital IC's	95p
226	How to Build Advanced Short Wave Receivers	1.20p
227	Beginners Guide to Building Electronic Projects	1.25p
228	Essential Theory for the Electronics Hobbyist	1.25p
RCC	Resistor Colour Code Disc Calculator	10p